AF410706

Running Dynamics of Rail Vehicles

Running Dynamics of Rail Vehicles

Editors

Larysa Neduzha
Jan Kalivoda

MDPI • Basel • Beijing • Wuhan • Barcelona • Belgrade • Manchester • Tokyo • Cluj • Tianjin

Editors
Larysa Neduzha
Ukrainian State University of
Science and Technologies
Ukraine

Jan Kalivoda
Czech Technical University in
Prague
Czech Republic

Editorial Office
MDPI
St. Alban-Anlage 66
4052 Basel, Switzerland

This is a reprint of articles from the Special Issue published online in the open access journal *Energies* (ISSN 1996-1073) (available at: https://www.mdpi.com/journal/energies/special_issues/RunningDynamics_Vehicles).

For citation purposes, cite each article independently as indicated on the article page online and as indicated below:

LastName, A.A.; LastName, B.B.; LastName, C.C. Article Title. *Journal Name* **Year**, *Volume Number*, Page Range.

ISBN 978-3-0365-5379-5 (Hbk)
ISBN 978-3-0365-5380-1 (PDF)

© 2022 by the authors. Articles in this book are Open Access and distributed under the Creative Commons Attribution (CC BY) license, which allows users to download, copy and build upon published articles, as long as the author and publisher are properly credited, which ensures maximum dissemination and a wider impact of our publications.
The book as a whole is distributed by MDPI under the terms and conditions of the Creative Commons license CC BY-NC-ND.

Contents

Jan Kalivoda and Larysa Neduzha
Running Dynamics of Rail Vehicles
Reprinted from: *Energies* **2022**, *15*, 5843, doi:10.3390/en15165843 . . . . **1**

**Olexandr Shavolkin, Iryna Shvedchykova, Juraj Gerlici, Kateryna Kravchenko and František
Pribilinec**
Use of Hybrid Photovoltaic Systems with a Storage Battery for the Remote Objects of Railway
Transport Infrastructure
Reprinted from: *Energies* **2022**, *15*, 4883, doi:10.3390/en15134883 . . . . **5**

Krzysztof Polak and Jarosław Korzeb
Acoustic Signature and Impact of High-Speed Railway Vehicles in the Vicinity of Transport
Routes
Reprinted from: *Energies* **2022**, *15*, 3244, doi:10.3390/en15093244 . . . . **25**

**Sergey Goolak, Viktor Tkachenko, Svitlana Sapronova, Vaidas Lukoševičius, Robertas
Keršys, Rolandas Makaras, Artūras Keršys and Borys Liubarskyi**
Synthesis of the Current Controller of the Vector Control System for Asynchronous Traction
Drive of Electric Locomotives
Reprinted from: *Energies* **2022**, *15*, 2374, doi:10.3390/en15072374 . . . . **45**

**Włodzimierz Idczak, Tomasz Lewandrowski, Dominik Pokropski, Tomasz Rudnicki and
Jacek Trzmiel**
Dynamic Impact of a Rail Vehicle on a Rail Infrastructure with Particular Focus on the
Phenomenon of Threshold Effect
Reprinted from: *Energies* **2022**, *15*, 2119, doi:10.3390/en15062119 . . . . **65**

Ka Zhang, Jianwei Yang, Changdong Liu, Jinhai Wang and Dechen Yao
Dynamic Characteristics of a Traction Drive System in High-Speed Train Based on
Electromechanical Coupling Modeling under Variable Conditions
Reprinted from: *Energies* **2022**, *15*, 1002, doi:10.3390/en15031202 . . . . **85**

Jerzy Kisilowski and Rafał Kowalik
Mechanical Wear Contact between the Wheel and Rail on a Turnout with Variable Stiffness
Reprinted from: *Energies* **2021**, *14*, 7520, doi:10.3390/en14227520 . . . . **105**

Editorial

Running Dynamics of Rail Vehicles

Jan Kalivoda [1],* and Larysa Neduzha [2]

[1] Faculty of Mechanical Engineering, Czech Technical University in Prague, Technicka 4, 160 00 Prague, Czech Republic
[2] Faculty of Transport Engineering, Ukrainian State University of Science and Technologies, Lazaryan St. 2, 49010 Dnipro, Ukraine
* Correspondence: jan.kalivoda@fs.cvut.cz; Tel.: +420-224-352-493

Citation: Kalivoda, J.; Neduzha, L. Running Dynamics of Rail Vehicles. *Energies* **2022**, *15*, 5843. https://doi.org/10.3390/en15165843

Received: 19 July 2022
Accepted: 9 August 2022
Published: 11 August 2022

Publisher's Note: MDPI stays neutral with regard to jurisdictional claims in published maps and institutional affiliations.

Copyright: © 2022 by the authors. Licensee MDPI, Basel, Switzerland. This article is an open access article distributed under the terms and conditions of the Creative Commons Attribution (CC BY) license (https://creativecommons.org/licenses/by/4.0/).

1. Introduction

The importance of simulation calculations in developing railway vehicles and their subsystems is consistently growing compared to physical experiments and track tests. Computer simulations of vehicle running dynamics are gaining an increasingly important role in the railway vehicle acceptance process, the development of innovative running gears, the development of active controlled suspension systems, the prediction of wear of wheels and rails, the prediction of noise emissions, the optimization of energy consumption and many other current topics. The investigation of railway vehicles running dynamics plays an important role in many research and development areas of railway engineering, such as:

- Innovative running gear designs;
- Utilisation of active controlled systems in running gears;
- Wheel–rail force interaction;
- Wear of wheels and rails;
- Testing and simulation of the acceptance of running characteristics of railway vehicles;
- Verification and validation of simulation models;
- Vehicle dynamics modelling and simulation;
- Co-simulation and model coupling;
- Reduction in the noise generated by railway transport;
- Reduction in the energy consumption of railway transport;
- Alternative energy sources;
- Rail transport safety.

The purpose of this Special Issue is to publish the latest developments and results of theoretical and experimental research and to show the current problems their solutions in the simulation and testing of the running dynamics of railway vehicles.

2. A Short Review of the Contributions in This Issue

The call for papers to this Special Issue generated 12 submissions of very interesting, high-quality papers from Poland, Ukraine, Czech Republic, Slovakia, China and Lithuania. Five papers were published.

Kisilowski and Kowalik focused on the wear of turnout components of high-speed tracks. In their work [1], the simulation results of a high-speed train passing a turnout under variable conditions are presented, and dynamic forces acting in wheel–rail contacts and the wear coefficient are evaluated. The variable track stiffness along the switch is taken into the account in their model. The wear process of turnout elements and wheels is presented.

Zhang et al. focused on high-speed train traction drives. The paper [2] describes a mathematical model considering the interaction of the gear pair, transmission system, circuit of the traction motor, and the direct torque control strategy. The results of the co-simulation of mechanical and electrical parts of the traction system under traction, constant speed, and braking conditions are presented and discussed.

Idczak et al. investigate the dynamic impact of rail transport on the surrounding infrastructure with a particular focus on the phenomenon of the threshold effect within the transition zones of an engineering facility [3]. The problem of locally variable stiffness of the railway infrastructure, which could lead to accelerated infrastructure degradation, is identified. Theoretical results are compared to the field measurements conducted on a real track dynamically loaded with various types of passing vehicles.

The paper of Goolak et al. [4] deals with the analysis of the operating conditions of traction drives of the electric locomotives with asynchronous traction motors. The method of current controller synthesis based on the Wiener–Hopf equation was proposed to enable the efficient performance of the traction drive control system under the stochastic fluctuation of the catenary system voltage. The simulation results of the performance of the proposed current controller are presented and compared to the performance of the current controller used in the existing vector control systems of the traction drives of electric locomotives.

Polak and Korzeb focus on the acoustic impact of a train travelling at the speed of 200 km/h in straight sections and track curves [5]. The field test results were applied to the construction and verification of the model of sound propagation. The results of the test field measurements are presented, the main sources of noise coming from the studied train unit are identified and the dominant amplitude–frequency are determined.

The energy that an electrified railway line takes from the electrical grid fluctuates significantly, and the peaks often reach the power limit of a grid. In [6], Olexandr Shavolkin et al. propose to solve this issue using a photovoltaic system and battery storage working in parallel to the electrical grid supply. The mathematical model of such electrical system has been developed; based on the simulations of the parameters of the photovoltaic system, they have been justified; and the system's capabilities in the daily mode for different seasons of the year have been assessed.

3. Conclusions

The investigation of rail vehicle running dynamics plays an important role in the more than 200 year development of railway vehicles and infrastructure. Currently, there are a number of new requirements for rail transport associated with the reduced environmental impact, energy consumption and wear, whilst increasing train speed and passenger comfort. Therefore, the running dynamics of rail vehicles is still a research topic that requires improved simulation tools and experimental procedures. We would like to thank the authors of the papers included in the Special Issue "Running Dynamics of Rail Vehicles" in the journal *Energies* for their contribution to this topic.

Author Contributions: The authors contributed equally to this work. All authors have read and agreed to the published version of the manuscript.

Funding: This editorial received no external funding.

Acknowledgments: We thank the authors of all papers published in this Special Issue for submitting their excellent works. Additionally, we would like to thank all the reviewers for their recommendations and comments, which have improved the quality of this Special Issue. We are also grateful to MDPI for the invitation to act as Guest Editors of this Special Issue and to the Editorial Board of "*Energies*" for their kind co-operation, patience, and involvement in the preparation of this Special Issue.

Conflicts of Interest: The authors declare no conflict of interest.

References

1. Kisilowski, J.; Kowalik, R. Mechanical Wear Contact between the Wheel and Rail on a Turnout with Variable Stiffness. *Energies* **2021**, *14*, 7520. [CrossRef]
2. Zhang, K.; Yang, J.; Liu, C.; Wang, J.; Yao, D. Dynamic Characteristics of a Traction Drive System in High-Speed Train Based on Electromechanical Coupling Modeling under Variable Conditions. *Energies* **2022**, *15*, 1202. [CrossRef]
3. Idczak, W.; Lewandrowski, T.; Pokropski, D.; Rudnicki, T.; Trzmiel, J. Dynamic Impact of a Rail Vehicle on a Rail Infrastructure with Particular Focus on the Phenomenon of Threshold Effect. *Energies* **2022**, *15*, 2119. [CrossRef]
4. Goolak, S.; Tkachenko, V.; Sapronova, S.; Lukoševičius, V.; Keršys, R.; Makaras, R.; Keršys, A.; Liubarskyi, B. Synthesis of the Current Controller of the Vector Control System for Asynchronous Traction Drive of Electric Locomotives. *Energies* **2022**, *15*, 2374. [CrossRef]
5. Polak, K.; Korzeb, J. Acoustic Signature and Impact of High-Speed Railway Vehicles in the Vicinity of Transport Routes. *Energies* **2022**, *15*, 3244. [CrossRef]
6. Shavolkin, O.; Shvedchykova, I.; Gerlici, J.; Kravchenko, K.; Pribilinec, F. Use of Hybrid Photovoltaic Systems with a Storage Battery for the Remote Objects of Railway Transport Infrastructure. *Energies* **2022**, *15*, 4883. [CrossRef]

Article

Use of Hybrid Photovoltaic Systems with a Storage Battery for the Remote Objects of Railway Transport Infrastructure

Olexandr Shavolkin [1], Iryna Shvedchykova [1], Juraj Gerlici [2], Kateryna Kravchenko [2,*] and František Pribilinec [2]

[1] Department of Computer Engineering and Electromechanics, Institute of Engineering and Information Technologies, Kyiv National University of Technologies and Design, Nemyrovycha-Danchenka Street, 2, 01011 Kyiv, Ukraine; shavolkin.oo@knutd.edu.ua (O.S.); shvedchykova.io@knutd.edu.ua (I.S.)

[2] Department of Transport and Handling Machines, Faculty of Mechanical Engineering, University of Žilina, Univerzitná 8215/1, 0102632 Zilina, Slovakia; juraj.gerlici@fstroj.uniza.sk (J.G.); frantisek.pribilinec@fstroj.uniza.sk (F.P.)

[*] Correspondence: kateryna.kravchenko@fstroj.uniza.sk; Tel.: +421-41-513-2660

Citation: Shavolkin, O.; Shvedchykova, I.; Gerlici, J.; Kravchenko, K.; Pribilinec, F. Use of Hybrid Photovoltaic Systems with a Storage Battery for the Remote Objects of Railway Transport Infrastructure. *Energies* **2022**, *15*, 4883. https://doi.org/10.3390/en15134883

Academic Editors: Larysa Neduzha and Jan Kalivoda

Received: 25 May 2022
Accepted: 30 June 2022
Published: 2 July 2022

Publisher's Note: MDPI stays neutral with regard to jurisdictional claims in published maps and institutional affiliations.

Copyright: © 2022 by the authors. Licensee MDPI, Basel, Switzerland. This article is an open access article distributed under the terms and conditions of the Creative Commons Attribution (CC BY) license (https:// creativecommons.org/licenses/by/ 4.0/).

Abstract: The use of a grid-tied photovoltaic system with a storage battery to increase the power of objects of railway transport infrastructure above the limit on consumption from the grid with the possibility of energy saving is considered. The methods of analysis of energy processes in photovoltaic systems with a storage battery are used. They are added via the processing of archival data of power generation of a photovoltaic battery and computer modeling results. A technique of system parameter calculation to increase the power according to the given load schedule of the object at constant and maximum possible degree of power increasing is developed. The values of the average monthly generation of a photovoltaic battery at the location point of the object based on archival data are used. The principle of the control of power, consumed from the grid, according to the given values of the added and total load is developed. Using the basic schedule of added load power in connection with the graph of photovoltaic battery generation allows reducing the installed power of the storage battery. The additional reduction in the installed power of the photovoltaic and storage batteries is possible at the corresponding choice of the degree of power load increasing. The joint formation of current schedules with reference to the added power value and state of charge of the battery according to the short-term forecast of the generation of a photovoltaic battery is proposed. The value of added power at certain intervals of time is set according to the graph of actual generation of the photovoltaic battery, which contributes to the maximum use of its energy. With the average monthly generation of a photovoltaic battery in the spring–autumn period, the discharge of the battery during the hours of the morning load peak is not used. This reduces the number of deep discharge cycles and extends the battery life. The description of energy processes in steady-state conditions for the daily cycle of system functioning is formalized. On this basis, a mathematical model is developed in MATLAB with an estimation of the costs of electricity consumed from the grid. When modeling, archival data are used for days when the generation of a photovoltaic battery over time intervals is close to average monthly values. This makes it possible to evaluate the effectiveness of system management under conditions close to real during the year.

Keywords: energy saving; control by forecast; power limit of consumption; railway transport infrastructure object; simulation in the daily cycle

1. Introduction

Photovoltaic systems (PVSs) are the most common in the "green" energy sector. A considerable share of PVSs falls on local objects (LO) for various purposes, where hybrid solar power plants with connection to the alternating current distribution grid (DG) are used. This corresponds to the current trend of localizing consumption at the generation site [1].

The issue of localization of the consumption of energy, generated by PV, can be solved when using a PVS to increase the load power of the LO above the limit on consumption from the grid. Such an application of PVS can be in demand during the development (expansion) of facilities with an increase in consumption, when the possibilities of increasing the capacity of the existing connection to the power grid are exhausted.

PVS application can be useful for railway infrastructure facilities remote from the transformer substation, including those with the seasonal nature of the load (consumption). The simplified structure of the power supply of LO with a PVS and storage battery (SB) is shown in Figure 1. The implementation of this option may be cheaper than laying a new power transmission line and replacing DG equipment. The advantage of this solution is the possibility of reducing the cost of paying for consumption from the DG during the term of operation of the PVS and the possibility of autonomous operation in the case of DG disruptions.

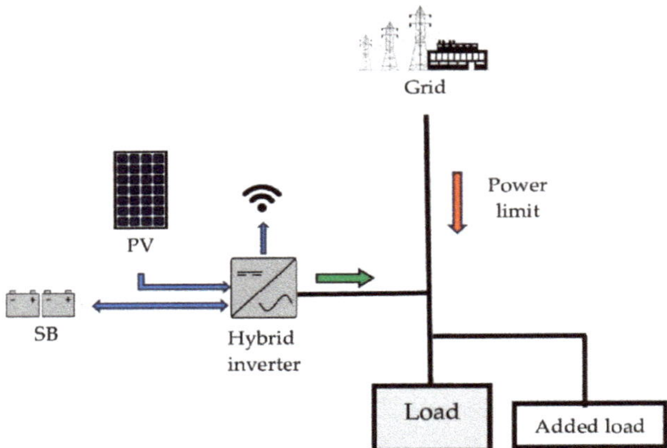

Figure 1. Simplified structure of power supply of LO load.

This also applies to objects of extensive railway transport infrastructure, including remote objects on sections of the grid that are not currently electrified. The main task when using PVS for such facilities is to provide for their own needs. At the same time, the possibility of increasing their power when using existing DG is limited by the consumption limit. The issue of energy saving remains relevant. First of all, this concerns the reduction in electricity consumption from DG. The use of storage batteries in the PVS with modern methods of energy management [2] allows rationally redistributing the energy in the system in time. This also achieves a reduction in consumption from DGs and an increase in the reliability of the power supply of LO. Improving the performance of systems with renewable sources of electricity is an urgent task, which contributes to the further development of energy with distributed sources of electricity.

2. Literature Review and Problem Statement

The demand for the use of hybrid PVS with SB is confirmed by the fact that the electrical market is widely represented by various solutions of hybrid inverters [3,4], which are designed for LO. They have all the equipment for connecting a photovoltaic battery (PV) and SB, as well as sufficiently powerful software. They are designed for self-consumption of LO while reducing consumption from DG, and they provide an uninterruptible power supply function.

With a power consumption of LO in excess of 10 kW, it is advisable to use three-phase multifunctional inverters while maintaining a power factor close to 1 at the point of common coupling (PCC) to the grid [5–12]. This allows to unload the grid from reactive power and ensures the symmetry of loading the phases of the grid at an unbalanced load [9–12]. A previous study [13] presented a three-phase PVS with an SB with four converters with a common link of direct current (DC). In the DC link, a proportional–integral (PI) voltage controller (VC) was used. The controller set the currents of the battery and supercapacitor. The "multiconverter" provided a given active power with the proper quality of electricity, PV operation with tracking the maximum power point (MPPT), and extended battery life with hybrid storage. In [10,11], the structure of the control system of a multifunctional inverter of a PVS with an SB with voltage stabilization in the DC link with three VCs was considered. The VC controls the currents of the SB and PV, as well as current in the PCC. The structure changes in accordance with the mode of operation, and only one of the VCs is always used. Ensuring the efficiency of the use of hybrid PVS with energy storage for LO is usually associated with a reduction in the cost of consuming electricity from the grid and an increase in the reliability of the power supply [14]. With a wide range of changes in PV generation during the year, cost reduction is achieved by overestimating the power of PV relative to the load power. This allows providing acceptable indicators in cloudy weather and winter.

PV generation changes significantly during the day and year. Therefore, the increase in the efficiency of PVS energy management is associated with the use of the forecast of PV generation [15–19]. The issues with obtaining an accurate forecast were considered in a number of studies, particularly [15,16]. Recently, web resources have been made available that provide a forecast with a discreteness of 0.5 h or less, including an individual one at the location [20,21]. For the application without generation to the grid, the main purpose of the forecast is load planning for the next day and the possibility of adjustment when it changes [19].

Certain opportunities to reduce the cost of paying for electricity consumed by LO from DG are provided by taking into account the tariffication of payment [19,22–25]. Features of the implementation of PVS in the application of static and dynamic tariffs were considered in [22].

When using PV for the needs of LO (without generating electricity to the grid), it is achievable to reduce the cost of electricity consumption from DG by up to five times at one tariff rate (up to seven times at wo rates) in the summer [19,26]. In winter, the reduction is insignificant—around 1.2 times. A general estimation of cost reduction for the year is not given. In [27], along with meeting the needs of the LO, the use of the planned generation of electricity in the grid during peak hours was considered. This allowed significantly reducing electricity costs. Overall estimation of cost reduction during the year was not given. Regardless, there was still an underutilization of PV energy in the summer.

An effective tool for assessing the efficiency of the energy management of a PVS with an SB is mathematical modeling [26–29]. Modeling of a hybrid system with a supercapacitor for the PV generation period was considered in [29]. Modeling of energy processes in the daily cycle with an estimate of the cost of paying for electricity consumed from the DG was considered in [26,27]. The use of archival data on PV generation [30] allowed studying the operation of the system in different weather conditions with an estimation of the cost of electricity, consumed from the DG.

The possibilities of PVS use for power increase for infrastructure objects, remote from the transformer substation, over the limit of consumption have not been sufficiently studied. This is related to the determination of parameters and limiting capabilities in different seasons of the year, the features of the formation of the SB state of charge in the process of operation, and the realization of the principles of added power formation at maximum use of PV energy. In this case, it is possible to reduce the installed power of the PV and the battery. An important role in assessing the capabilities of the system is performed by mathematical modeling.

Thus, the purpose of the article is to develop principles for the use of PVS with SB to increase the power of the LO above the power limit for consumption from the grid with the maximum use of PV generation.

The main objectives of the research are as follows:

- to study the possibilities of increasing the load power of the LO above the power limit for consumption from the grid during the year;
- to justify the choice of PVS parameters with the formation of the graph of added power according to the accepted load schedule of the LO with a decrease in the installed power of the PV and battery;
- to develop principles for the implementation of the management of PVS using the forecast of PV generation;
- to perform an assessment of the system's capabilities in the daily mode for different seasons of the year using mathematical modeling.

3. Methodology of Research

A study of ways to improve the control mechanism of the PVS with an SB for increasing the power of the LO was carried out on the basis of analytical methods in electrical circuits. The results of processing statistical data on the PV generation for a given point of location of the object were also used. A proportional increase in power to the original load schedule was adopted. As original, the load schedule characteristic of objects with a predominance of day loads, with peak loads in the morning and evening and a decrease in the load at night, was considered. The basic schedule of power, added and provided by the PVS, was adopted in accordance with the PV generation schedule. The maximum value of the power increase factor in winter takes into account the possibility of ensuring the battery charge within the power limit for consumption from the grid. On this basis, the energy capacity of the battery was determined, followed by an assessment of the possibilities for increasing the load power, taking into account the average monthly PV generation. The choice of a fixed value of the degree of power increase was carried out while taking into account the use of PV energy and cost reduction. The control system of the PVS converter unit was implemented on the basis of a classic double-loop structure with voltage stabilization in the DC link. When forming the $SOC(t)$ (state of charge) of the SB schedule, a limit of DOD $\leq 80\%$ (depth of discharge) was introduced with one deep discharge per day in the spring–summer–autumn period. The technique to calculate the reference of added power for maximum use of PV energy was realized on the basis of an analysis of the average monthly PV generation by time intervals for the taken load schedule. Analysis of energy processes in the system "DG–PVS with SB–LO load" was carried out for the daily cycle without taking into account transient processes and higher harmonics in energy converters. Energy losses were accounted for through efficiency. The properties of the SB were considered in accordance with the characteristics of the manufacturer. PV generation was estimated on the basis of monthly average values for given time intervals during the day. Data of PV generation were obtained for the location point of the object when processing archival data for 5 years. The modeling of energy processes was performed using MATLAB software package using real archive graphs of PV generation. The days were chosen when PV generation by time intervals was close to the average monthly values. The model was completed on the basis of analytical expressions for steady-state operating modes, which correspond to generally accepted proven calculation methods. When the operating modes of the system changed, the corresponding calculated expressions were used at time intervals per day.

4. The Results of the Research on the Use of PVS with SB to Increase the Power of LO

The option of PVS with an SB with a grid multifunctional inverter VSI (Figure 2) was considered. The middle pin (n) of the VSI link DC was connected to the neutral connection point of the DG. This allowed ensuring the equalization of power consumption from the grid by phases under unbalanced load LO [9–11]. This also made it possible to control the active power P_g in the PCC. The structure of the PVS included the following elements: a DC voltage converter PV (CPV) with a transistor key for measuring the current of the PV short-circuit [10], and a DC voltage converter battery (CSB). The converter unit was controlled by a control unit (CU) connected to the programmable control unit (PCU). Communication with the web resource to obtain forecast data was provided by a Wi-Fi WFM module.

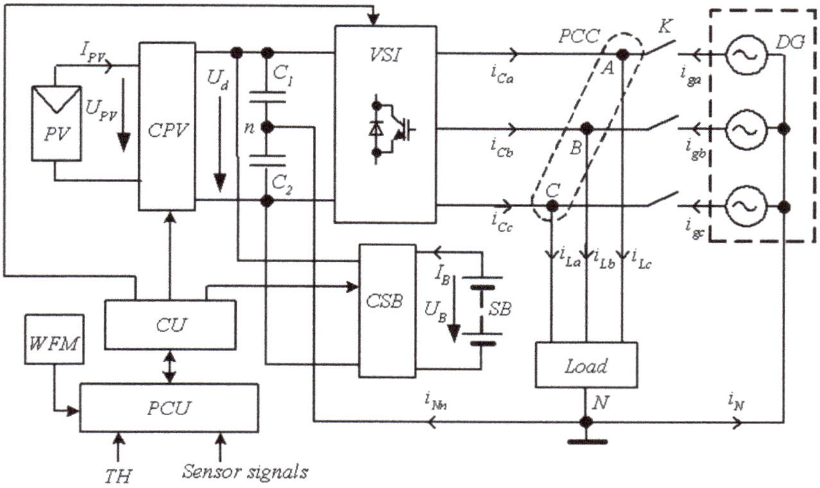

Figure 2. Structure of hybrid PVS with SB.

When solving the issue of increasing the power of the LO, the general approach changes somewhat; the DG becomes an auxiliary source of energy of limited power. As originally proposed, the use of a load schedule was considered in accordance with the standard distribution of peak loads [19,26] for objects of the utility sector and the nondomestic sector with a single-shift mode of operation. The following distribution of load intervals was accepted: night tariff zone in the period May–August ($t_7 = 24.00$, $t_1 = 7.00$), day tariff ($t_1 = 7.00$, $t_2 = 8.00$), ($t_3 = 11.00$, $t_5 = 20.00$), ($t_6 = 23.00$, $t_7 = 24.00$), peak load zones ($t_2 = 8.00$, $t_3 = 11.00$), and ($t_5 = 20.00$, $t_6 = 23.00$); in the period autumn–winter–spring ($t_1 = 7.00$, $t_2 = 8.00$, $t_3 = 10.00$, $t_5 = 17.00$ and 18.00, $t_6 = 17.00$ and 18.00, $t_7 = 24.00$). An additional point of time t_4, was also used, corresponding to the transition to an evening decrease in PV generation. This time during the year varied from $t_4 = 16.00$ in June to $t_4 = 14.00$ in December.

Increasing the power of the LO load in the daytime assumes that the total power P_{LC} of the LO load is defined as $P_{LC} = P_{Lg} + P_C$ (where P_{Lg} is the power which is provided by consumption from the grid (P_{Lg} does not exceed the limit on consumption P_{LIM}), and P_C is the added power, generated by the inverter due to the energy of PV and SB). We can take $P_{Lg} = P_{LIM}$; then, with an increase in P_L; the possible value of P_{LC} grows with constant consumption from the grid. If the actual load power is less than P_{LC}, the electricity consumption from the grid is reduced.

The value of the energy generated by PV during the year varies widely (Table 1). Table 1 shows the data [30] on the average monthly generation of PV per day W_{PVAVD} at power $P_{PVR} = 1$ kW for the Kyiv location: latitude (decimal degrees)—50.451, longitude (decimal degrees)—30.524. Similar data are given for the city of Žilina W_{PVAVDZ} (latitude (decimal degrees)—49.224, longitude (decimal degrees)—18.748). In winter, the PV generation in Žilina is slightly higher. In Table 1 (in parentheses), monthly energy values per day (W^*_{PVAVD}) and at time intervals during the day (W_{PV23}, W_{PV34}, and W_{PV45}) are also presented for Kyiv. The energy values without parentheses correspond to the selected days when the generation was close to the monthly average. These values were obtained from archival data of P_{PV} generation in Kyiv [30] for the period 2012–2016.

Table 1. Average monthly PV generation during the year.

January	February	March	April	May	June	July	August	September	October	November	December
Average monthly generation of PV per day for Kyiv location W_{PVAVD}, kWh											
1.17	1.81	2.87	3.88	4.27	4.43	4.38	4.24	3.85	2.57	1.14	0.91
Average monthly generation of PV per day for Žilina location W_{PVAVDZ}, kWh											
1.272	2.053	2.811	3.799	3.76	3.914	4.073	3.84	3.434	2.6	1.418	1.25
Energy values for selected days when the PV generation was close to the average monthly generation per day in Kyiv (PV generation calculated using archival data for the period 2012–2016) W^*_{PVAVD}, kWh											
0.85 (0.98)	1.81 (1.81)	2.57 (2.83)	3.89 (3.83)	4.25 (4.26)	4.47 (4.48)	4.36 (4.37)	4.036 (4.16)	3.4 (3.52)	2.44 (2.49)	0.798 (0.96)	1 (1.03)
Energy values for selected days when the PV generation is close to the average generation at time interval (t_2, t_3) in Kyiv (PV generation calculated using archival data for the period 2012–2016) W_{PV23}, kWh											
0.144 (0.166)	0.31 (0.36)	0.5 (0.55)	0.62 (0.52)	1.06 (1.15)	1.26 (1.17)	1.17 (1.13)	0.581 (1.08)	0.41 (0.56)	0.34 (0.33)	0.225 (0.22)	0.166 (0.21)
Energy values for selected days when the PV generation is close to the average generation at time interval (t_3, t_4) in Kyiv (PV generation calculated using archival data for the period 2012–2016) W_{PV34}, kWh											
0.643 (0.78)	1.31 (1.36)	1.83 (1.84)	2.09 (2.38)	2.43 (2.43)	2.51 (2.27)	2.46 (2.5)	2.78 (2.45)	2.25 (2.27)	1.95 (1.74)	0.535 (0.69)	0.806 (0.8)
Energy values for selected days when the PV generation is close to the average generation at time interval (t_4, t_5) in Kyiv (PV generation calculated using archival data for the period 2012–2016) W_{PV45}, kWh											
0.061 (0.038)	0.16 (0.083)	0.22 (0.32)	0.79 (0.84)	0.539 (0.53)	0.62 (0.63)	0.57 (0.61)	0.66 (0.53)	0.65 (0.72)	0.148 (0.39)	0.036 (0.03)	0.03 (0.021)

The value of the added load power P_C was determined on the basis of the average monthly daily generation of $W_{PVAVD} \approx 2500$ W in the transitional seasons of the year (Table 1): October and March. In November–February, the load, provided by PV, decreased. The average value of power in the daytime ($P_{AVD} = W_{PVAVD}/t_D$, where t_D is the length of the day) was about 200 W.

We took the basic load schedule (average P_L values by time intervals) taking into account peak loads in the morning and evening with a decrease in load after t_4 until the evening peak, e.g., $P_{L23B} = 200$ W (P_{LAVD}), $P_{L34B} = 180$ W ($0.9\ P_{LAVD}$), $P_{L45B} = 160$ W ($0.8\ P_{LAVD}$), and $P_{L56B} = 200$ W (P_{LAVD}). The total night load of LO could be taken from the condition $P_{Lg62} = P_{LIM} - P_B$ (we took P_{LIM} equal to the peak power $P_{LIM} = P_{L23B} = 200$ W, $P_B = U_B I_B$—power, consumed from the grid to charge the SB (U_B and I_B—voltage and current of the SB)). At the same time, $P_{Lg62} \geq P_{LMIN}$, $P_{LgMIN} = 0.2 P_{L56}$ in summer, and $P_{LgMIN} = 0.3 P_{L56}$ in winter. The total energy transmitted by the inverter to increasing power (P_L) at the interval (t_2, t_6) was $W_{L26} = 2740$ Wh in summer, $W_{L26} = 2580$ Wh in autumn–spring, and $W_{L26} = 2410$ Wh in winter.

The effective use of the SB's capabilities for the redistribution of energy in the system involves the formation of the $SOC(t)$ dependency $Q^*(t)$ ($Q^* = 100Q/Q_R$, $Q = Q_0 + \int I_B dt$, $Q_R = C_B$—the rated value (100%) or capacity (Ah) of the battery, Q_0—the initial value). Increasing the power during peak hours in the morning and evening, when the PV generation is small, implies a deep discharge of the SB. That is, we have two deep discharges per day. In these conditions, the use of lithium-ion batteries is preferable.

Two variants of implementation were considered: (1) with a maximum increase in power in accordance with PV generation, which is available for the consumers with seasonal load; (2) with a constant increase in power during the year.

For variant (1), it was assumed that the planning of load using the day-ahead forecast was possible.

For calculation of the value of SB energy capacity, in the interval (t_4, t_6), the PV generation W_{PV45} is small, and the increase in power is achieved mainly due to the energy of the SB. At the same time, the energy balance is determined by the following expression:

$$0.01 \cdot \Delta Q^*_{46} W_B \cdot \eta_C \cdot \eta_B = P_{C45}(t_5 - t_4) + P_{C56}(t_6 - t_5) - W_{PV45} \cdot \eta_C, \tag{1}$$

where W_B is the SB energy capacity ($W_B = U_B C_B$), $\Delta Q^*_{46} = Q^*_4 - Q^*_6$, η_C is the general efficiency of the SB voltage converter and grid inverter, and η_B is the efficiency of the SB.

W_B is calculated from the condition of the functioning of the added power of the load during the evening peak hours (t_5, t_6) and on the intervals (t_4, t_5), when PV generation is significantly reduced. This is most typical in winter, with a longer evening peak (4 h). The limitation is to ensure the possibility of the battery charge and the operation of the load at night within the framework of P_{LIM}. When $W_{PV46} = 0$, the value of the energy capacity of the battery SB is

$$W_B = \frac{W_{C46}}{0.01 \cdot \Delta Q^*_{46} \cdot \eta_C \cdot \eta_B}, \tag{2}$$

where $W_{C46} = W_{L46}(\rho - 1)$; $\rho > 1$ indicates the degree of increase in the load power.

It was assumed that the night load in winter also proportionally increases $P_{Lg} = \rho\, 0.3\, P_{LIM}$. The possible value for the battery charge in the framework of limit is denoted as

$$\Delta W_{B62} = (t_2 - t_6)P_{LIM}(1 - 0.3\rho). \tag{3}$$

Alternatively,

$$\Delta W_{B62} = \frac{W_{L46}(\rho - 1)\Delta Q^*_{62}}{\Delta Q^*_{46}(\eta_C \cdot \eta_B)^2}. \tag{4}$$

The maximum value ρ_{MAX} in the interval (t_4, t_6) can be determined in accordance with Equations (2) and (3). In this case, at $DOD_6 \leq 80\%$, $\rho_{MAX} = 1.721$. Accepting $\rho = 1.7$, $W_B = 1164$ Wh. At $DOD_6 \leq 90\%$, the value is $W_B = 1034$ Wh. The average value $W_B = 1099$ Wh (corresponding to, for example, $C_B = 43$ Ah at $U_B = 25.6$ V) can be accepted. The resulting value is sufficient to ensure the added power at $\rho_{MAX} = 1.7$ and average monthly PV generation in December in the interval (t_2, t_6).

However, for the same value $P_{LC} = \rho P_L$ and $P_{Lg} \leq P_{LIM}$, the different variants of reference of the added power P_C for the interval (t_2, t_6) (in Table 2, data are presented for $\rho = 1.7$) are possible. This allows planning $P_C(t)$ according to the conditions.

Table 2. Variants of the reference P_C and P_{Lg} at $\rho = 1.7$.

	Interval	(t_2, t_3)	(t_3, t_4)	(t_4, t_5)	(t_5, t_6)
Variant	P_{LC}, W	340	306	272	340
va	P_C, W	140	126	112	140
	P_{Lg}, W	200	180	160	140
vb	P_C, W	140	106	72	140
	P_{Lg}, W	200	200	200	200
vc	P_C, W	140	200	72	140
	P_{Lg}, W	200	106	200	200

Above the variant, *va* was considered. At minimal PV generation (in winter), variant *vb* ensured the biggest value ρ, whereby $P_{C26}(t) = \rho P_{L26}(t) - P_{LIM}$. In this case, $W_{C46} = \rho W_{L46} - P_{LIM}(t_6 - t_4)$, and the value ΔW_{B62} can be expressed as

$$\Delta W_{B62} = \frac{P_{LIM}(6\rho - (t_6 - t_4))\Delta Q^*_{62}}{\Delta Q^*_{46}(\eta_C \cdot \eta_B)^2}. \tag{5}$$

We get values $\rho_{MAX} = 1.78$ and $W_B = 968$ Wh (corresponding to, for example, $C_B = 37.8$ Ah at $U_B = 25.6$ V). The advantage of this variant is the minimal value of P_C in the interval (t_4, t_6) when the PV generation is minimal.

Variant *vc* is included in Table 2. This variant is more tied to the PV generation schedule during the day.

There are days in winter when $W_{PVD} \leq 60$ Wh. In this case, it is possible to use a night charge of the battery up to 100%, followed by a discharge during the day (from 8:00 a.m. to 9:00 p.m.) at $DOD_6 = 10\%$. This allows using the energy $\Delta W_{B26} = 768$ Wh (at $W_B = 968$ Wh) in the load. Accordingly, $\rho = (\Delta W_{B26} + W_{L26LIM})/W_{L26LIM} = 1.3$. At small values of W_{PVD} (for example, at $W_{PVD} \leq 300$ Wh), there is the possibility to redistribute the energy of SB by intervals. For example, the degree of increase in power in peak hours can be saved without increasing from 10:00 a.m. to 5:00 p.m. Then, for example, $0.3\,\Delta W_{B26}$ can be used in the morning peak, and, taking into account the duration, $0.6\Delta W_{B26}$ can be used in the evening peak. This allows ensuring $\rho \approx 1.6$ in intervals (t_2, t_3) and (t_5, t_6).

In the period spring–autumn, there is the possibility to increase ρ. Determination of ρ is carried out in the accordance with forecast data of PV generation on the next day on the basis of the balance of energy in the intervals (t_2, t_6) and (t_4, t_6).

It is possible to calculate ρ for the interval (t_2, t_6) as follows:

$$\rho_{26} = \frac{W_{PV26}\eta_C + \Delta W_{B26} - W_{gR26} + P_{LIM26}}{W_{L26}}, \tag{6}$$

where W_{gR26} indicates a decrease in energy consumption from the grid, and $W_{LIM26} = P_{LIM}(t_6 - t_2)$.

Depending on the PV generation, the implementation is possible (a) with the discharge of the SB without a decrease in consumption from the grid, (b) without the discharge of the SB and a decrease in consumption from the grid, (c) with minimal discharge of the SB and a decrease in consumption from the grid, or (d) without the discharge of the SB and a decrease in consumption from the grid. Variant (a) is specific to winter with small PV generation, when $\rho_{MAX} = 1.7$ is accepted.

For interval (t_4, t_6),

$$\rho_{46} = \frac{kW_{PV46}\eta_C + \Delta W_{B46} - W_{gR46} + P_{LIM46}}{W_{L46}}, \tag{7}$$

where k is a coefficient, taking into account the use of PV energy for consumption, and ΔW_{B46} corresponds to $DOD = 80\%$.

The value of k takes into account that, when the SB is fully charged by the time t_4, only part of the PV energy is used for consumption by the load. The rest of the energy provides a reduction in consumption from the grid. Thus, $k = 1$ is accepted if $W_{PV45} \cdot \eta_C < W_{C45}$, while $k = 0.5$ is accepted if $W_{PV45} \cdot \eta_C \geq W_{C45}$.

Furthermore, ρ_{46} can be defined at the decrease in consumption in Equation (6) when $W_{gR46} = 0$. For March (Table 1), $\rho_{46} = 1.8$. Compared with the value $\rho_{26} = 2.02$ according to Equation (6) for option (b), we accept the smaller value. In this case, with $\rho = 1.8$, it is possible to implement option (c). Thus, for $\rho = 1.8$, we find the following value:

$$\Delta Q^*_{23} = \frac{W_{PV23}\eta_C - W_{C23}}{0.01W_B\eta_C\eta_B}. \tag{8}$$

For March $\Delta Q^*_{23} = 17\% > 0$, i.e., the SB state of charge increases, and the night battery charge is not needed. Then, in accordance with Equation (6), we have a decrease in consumption on the order of 572 Wh. With the average monthly values of PV generation for the summer period, $\rho = 1.95$ is achievable.

A graph of the added power $P_C(t)$ and the state of charge of the SB $Q^*(t)$ is presented in Figure 3 in accordance with the obtained value of ρ. We consider options *vb* and *vc* when using the PV energy W_{PV} forecast data over time intervals. At the same time, it is desirable to (1) use the lowest possible power consumption at night (until 8:00 a.m.) per battery charge, and (2) ensure the condition $Q^*_4 \to 100\%$, taking into account the reduction in PV generation in the evening. A prerequisite is the maximum use of PV energy.

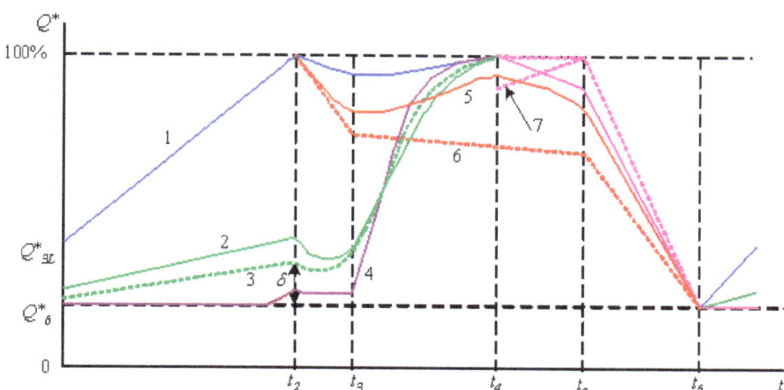

Figure 3. Formation of schedule of state of the charge of SB.

Reference to the initial value Q^*_{2R} is carried out in accordance with values ΔQ^*_{23} and ΔQ^*_{24} (calculated similarly to Equation (8)). If $\Delta Q^*_{24} \leq 0$, then $Q^*_{2R} = 100\%$ (curves 1, 5, and 6 in Figure 3). In this case, when $\Delta Q^*_{24} > 0$ and $\Delta Q^*_{23} \leq 0$, then $Q^*_{2R} = (100 - \Delta Q^*_{24}) \geq 40\%$ (curve 2 in Figure 3). If $\Delta Q^*_{24} > 0$, $\Delta Q^*_{23} > 0$, and $W_{PV23} \cdot \eta_C / W^1_{C23} < 1.5$ (W^1_{C23} is the value for the basic schedule with given ρ), then $Q^*_{2R} = (100 - \Delta Q^*_{24}) \geq (Q^*_6 + \delta)$ (curve 3 in Figure 3, $\delta = 10$–15%). In all cases, the reference of the added power is $P_{C23} = P^1_{C23}$.

If $\Delta Q^*_{24} > 0$, $\Delta Q^*_{23} > 0$, and $W_{PV23} \cdot \eta_C / W^1_{C23} \geq 1.5$, then $P_{LC23} \geq (P_{C23} = P_{PV23} \cdot \eta_C) \geq P^1_{C23}$, which is given in accordance with PV generation. In this case, the SB charge (curve 4 in Figure 3) is not carried out. It also does not use the battery charge at night, but there may be some battery charge before 8:00 a.m. (sunny morning).

The P_{C34} value in the interval (t_3, t_4) is $P_{C34} = \frac{W_{PV34}\eta_C - 0.01\Delta Q^*_{34}W_B\eta_C\eta_B}{(t_4 - t_3)} \leq P_{LC34}$, where ΔQ^*_{34} is defined using values Q^*_{2R} and ΔQ^*_{23}, as described above.

In the interval (t_4, t_5) with a large PV generation, the state of SB charge can increase (curve 7 in Figure 3), be unchanged ($\approx$100%), or decrease. The possibility of an increase is excluded by increasing the P_{C45}, which is carried out automatically. The formation of the degree of discharge in the interval (t_5, t_6) is carried out in the mode of regulation of the SB current.

The battery charge when $Q^* \geq Q^*_d = 90$–92% is reached at a constant value of the voltage [31]. In this case, the battery current is determined by the charge curve and is significantly reduced. Thus, the ability of the SB to receive energy is limited. Therefore, when $Q^* \geq Q^*_d$, the value of P_C is set according to the actual PV generation as $P_{LC} \geq P_C = (P_{PV} \cdot \eta_C - P_B) \geq P^1_C$. The limitation $P_{LC} \geq P_C$ is implemented by reducing the PV generation. The introduction of this mode allows ensuring more complete use of the PV energy, particularly when the calculated value of P_{C34} is less than required.

Figure 3 also shows a graph for the case when the PV energy is not enough to charge the battery up to 100% (curve 5) and for the case when the PV generation is close to 0 (curve 6). In these cases, the task is realized to support the load during peak hours.

Consider the variant with the constant increase of power during the year. For the choice of value ρ, consider the option vb with reference to the added power when the value W_B is minimal. According to Equations (2) and (5), when ρ increases, W_B also increases. An important issue is the underutilization of the installed PV power at high solar activity in the summer. To estimate, we introduce the PV energy utilization factor,

$$k_{PV} = \frac{W_{C26} + W_{gR26} - \Delta W_{B26}}{m_P W_{PV26}\eta_C},\tag{9}$$

where W_{gR26} indicates a decrease in energy consumption from the grid, W_{C26} is the energy consumed by the added load, and m_P is a coefficient of PV power recalculation relatively installed power $P_{PVR} = 1$ kW.

The value of W_{gR26} can take place during hours of high daytime solar activity t_{da}, when $P_{PV}.\eta_C > P_C + P_B$ ($P_B = U_B I_B$ is the power of SB charging). To exclude the generation of energy into the grid, the restriction $P_{gR} \leq P_{LIM}$ is used, which is achieved by regulating (reducing) the PV generation. In the limited case, $W_{gR26} = P_{LIM} \cdot t_{da}$ ($t_{da} = 6$ h, from 10:00 a.m. to 4:00 p.m.).

The value of ΔW_{B26} in the summer period with the average monthly PV generation is taken as equal to $\Delta W_{B26} = 0$ (nighttime battery charging is not required).

To assess the efficiency, we use the coefficient of cost reduction for electricity consumed from the grid (at one tariff rate and full use of PV energy). For the winter period (December), when the power increase in the interval (t_2, t_6) is achieved while maintaining the consumption from the grid within the limit,

$$k_E = \frac{W_{LC}}{W_g} = \frac{W_{LC}}{W_{LC} + \Delta W_{B26}/\left(\eta_C \eta_B\right)^2 - m_P W_{PV} \eta_C}, \tag{10}$$

where $W_{LC} = W_{LC62} + W_{C26} + P_{LIM}(t_6 - t_2)$ is the total energy consumed by the load.

At $m_P = 1$, with an increase in ρ and, accordingly, W_{C26}, the value of k_E decreases, and the value of k_P increases. If the installed power ($m_P < 1$) is reduced, then there is a reverse change in the coefficients.

Consider the option with $\rho = 1.6$ in comparison with the variant for $\rho = 1.7$ discussed above. This allows reducing the W_B value from 968 Wh to 800 Wh (by 21%). Almost the same value of k_E for December, in this case, can be obtained with $m_P = 0.86$ and an increase in k_P to 0.71 instead of 0.679 (in June). If we recalculate to the same load power, we get a decrease in the installed PV power by 9.4% and in the energy capacity W_B of the SB by 14%. At the same time, it remains possible to increase the power to $\rho = 1.8$.

The reference of the value of added power and the formation of a graph of the state of charge of the SB are carried out according to the method discussed above.

A simplified structure of the PVS control system is shown in Figure 4. The system of automatic control of the CU converter unit is implemented according to well-known principles with voltage stabilization in the DC link Ud at the VSI input [10,13]. This is provided by three proportional–integral (PI) voltage controllers (VCs): VCI$_{PV}$ forms the PV current reference; VCI$_B$ forms the SB current reference; VCIg forms the reference of the current at the point of common coupling to the grid. The system (Figure 4) contains three channels:

- PV generation control. This channel contains a control unit CPV (CSCPV), which provides the processing of the reference value of the current I^1_{PV}, current limiting unit LU1 with adjustable limit, and switch S2 for current settings in mode MPPT (position 2) or from VCI$_{PV}$ when regulating generation (position 1). Current limiting at the level $I_{PVM} = (0.9 \div 0.92)I_{SQ}$ excludes PV operation in the short-circuit mode [10];
- Charge control of SB. This channel contains a control unit CSB (CSCSB), which provides the processing of the reference value of the current I^1_B, current limiting unit LU3 charge and discharge of SB, and switch S3 for current settings from VCI$_B$ (position 2) or PCU (position 1);
- Grid current I_g control (reference of power consumed from the grid). This channel contains a reference current unit and an inverter current control loop i_C (RCU + CCL) and input of reference of the amplitude of the grid current (I^1_{gm}), which, via switch S1, connects to VCIg (position 2) or PCU (position 1). Limitation unit LU2 has a lower $I^1_{gm} \geq 0$, top I^1_{gmLIM} limits. Unit (RCU + CCL) provides i_g in PCC taking into account the phase currents of the inverter $i_{Ca,b,c}$ and load $i_{La,b,c}$ [10,11].

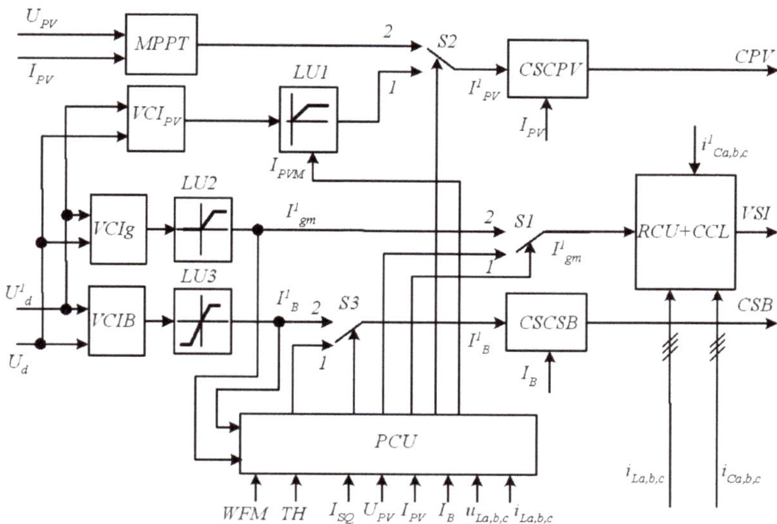

Figure 4. Structure of the control system of PVS.

The control of switches in the structure and the formation of current references are carried out by the PCU. The PCU also processes the prediction data according to the WFM block signal and the specified time intervals (TH). The phase lock loop (PLL) and offline control PVS channel when the mains voltage is turned off in Figure 4 are not shown (current reference of inverter phases $i^1_{Ca,b,c}$; in this case, a separate load voltage controller is set [10]).

The control principle (Table 3) is based on the control of active power, consumed from the grid (in PCC) $P_g = P_{LC} - P^1_C$ (P^1_C is the value of the added load power at the current time interval, and P_{LCi} is the value of the active load power according to the measured values of currents and phase voltages) at $P_{LIM} \geq P_g \geq 0$. The initial parameters are the recommended schedule (maximum average value) $P_{LCR}(t)$ and the base schedule $P_{CB}(t)$ for the accepted value of ρ, and the calculated schedule $P^1_C(t)$. There is a possible situation when $P_{LC} \neq P_{CR}$. With this in mind, deviation compensation $\Delta P_{LC} = (P_{LC} - P_{LCR}) \geq 0$ is used in the reference $P^1_C = P^1_C + \Delta P_{LC}$. The operating modes of the control system by intervals and the used controllers are given in Table 3. Calculation expressions for steady modes are given in the description of the model of energy processes.

Table 3. Operating modes of the PVS with SB.

Interval	(t_2, t_3)		(t_3, t_5)			(t_5, t_6)	(t_6, t_2)
Mode	*g1*	*g2*	*g3*	*g4*	*g5*	*g6*	*g7*
Reference I^1_{gm}	$P_g = P_{LC} - P_C,$ $P_{gi} \rightarrow I^1_{gm} \geq 0$	$P_g = P_{LC} - P_C,$ $P_g \rightarrow I^1_{gm} \geq 0,$	$P_g = P_{LC} - P_C,$ $P_g \rightarrow I^1_{gm} \geq 0,$	$VCI_g \rightarrow I^1_{gm} \geq 0$	$P_g = P_{LC} - P_C,$ $P_g \rightarrow I^1_{gm} \geq 0,$	$P_g = P_{LC} - P_C,$ $P_{gi} \rightarrow I^1_{gm} \geq 0$	$P_g = P_{LC} - P_C,$ $P_{gi} \rightarrow I^1_{gm} \geq 0$
Reference I^1_{PV}	MPPT	MPPT	$P_{PV}\eta_C \geq P_{LC} VCI_{PV} \rightarrow$ $I^1_{PV} I^1_{PV} \rightarrow P_{PVF}$	$P_{LC} > P_C \geq P_{CB},$ MPPT	$P_{CB} \geq P_C$MPPT	MPPT	MPPT
Reference I^1_B	$VCI_B \rightarrow I^1_B$	$VCI_B \rightarrow I^1_B$	$I^1_B = I_{BR} > I_B(Q)$ $I_B = I_B(Q)$	$I^1_B = I_{BR} > I_B(Q)$ $I_B = I_B(Q)$	$VCI_B \rightarrow I^1_B$	$VCI_B \rightarrow I^1_B$	$VCI_B \rightarrow I^1_B$
SOC	$Q^* \leq Q^*_d$	$Q^* \leq Q^*_d$	$Q^* > Q^*_d$	$Q^* > Q^*_d$	$Q^* \leq Q^*_d$	$Q^* < Q^*_d$	$Q^*_6 \rightarrow Q^*_2$

Consider the functioning of the system, starting from the intervals (t_2, t_3). PV operates in maximum power mode (MPPT). The SB current is set by the controller $VCI_B \rightarrow I^1_B$. Then, $I^1_{gm} = \sqrt{2}P_{gi}/3U_{gph}$ (U_{gph}—phase voltage of the grid) is defined according to $P_g = P_{LC} - P^1_{C23}$ and is supplied to the input of the current set unit RCU + CCL. When the load is reduced, P_g decreases; when the load is increased, P_g also increases. If $W_{PV23} \cdot \eta_C / W_{C23} \geq 1.5$, then the value of P_C is set by PV generation under the condition $P_{CL} \geq (P_C = P_{PV} \cdot \eta_C) \geq P^1{}_{C23}$. In this case, at $P_{PV} \cdot \eta_C < P^1{}_{C23}$, there is $P_C = P^1{}_{C23}$. At $P_{PV} \cdot \eta_C \geq P^1{}_{C23}$, there is $P_{PV}\eta_C = P_C$ and the current of SB charge decreases to 0; when $P_{PV}\eta_C \geq P_{LC}$, the SB charge is restored.

When switching to the interval (t_3, t_5) and $Q^* < Q^*{}_d$, the operating mode is saved. When $Q^* \geq Q^*{}_d = 90\text{–}92\%$, the SB current is determined by the charging characteristic $I_B(Q^*)$; if the set value $I^1{}_B > I_B(Q^*)$, then VCI$_B$ goes into saturation. The battery cannot consume all the energy. This leads to an increase in the voltage U_d on the capacitors in the DC link of the inverter. If $P_{LC} > P_C \geq P_{CB}$, then, upon reaching the switching threshold $(U_d + \Delta U)$, the VCI$_g$ controller ($g4$ mode) is activated and reduces $I^1{}_{gm}(P_g)$. If $P_{PV}\eta_C \geq P_{LC}$, then the VCI$_{PV}$ controller ($g3$ mode) is activated and $I^1{}_{PV}(P_{PV})$ is reduced. Thus, we have two conditions for switching of controllers (excluding the influence of transients).

In the interval (t_5, t_6), the discharge of SB is carried out with current (given by controller VCI$_B$).

$$I_{B56R} = \frac{0.01 C_B (Q^*{}_5 - Q^*{}_6)}{(t_6 - t_5)}.$$

The added power is set in accordance with an average value of SB voltage U_{BAV} as $P_{C56} = I_{B56R} \cdot U_{BAV} + \Delta P_{LC}$, if $\Delta P_{LC} = (P_{LC} - P_{LCR}) > 0$.

For the night period in the interval (t_6, t_2), the reference of the current of SB charge is

$$I_{B62R} = \frac{0.01 C_B (O^*{}_2 - Q^*{}_t)}{(t - t_2)},$$

where $Q^*{}_t$ is the current measured value, for example, with a resolution of 1 h.

In the period spring-autumn, the SB charge in the morning is possible from PV at $P_{PV} \cdot \eta_C \geq 0$.

Referencing of the value $Q^*{}_{2R}$ and planning of the recommended (maximum) load P_{LCR} is carried out according to the forecast for the next day. The given values of the added power in the intervals are specified at the beginning of the day in accordance with the current forecast of PV generation. Subsequently, with an interval of 1 h, the adjustment is carried out.

5. Modeling of Energy Processes in the Daily Cycle

The initial data for modeling were from an archive of PV generation on the mode of maximum power $P_{PVM}(t)$. Accordingly, the values of ρ, $Q^*{}_{2R}$, the base $P_{CB}(t)$ schedule for the accepted value of ρ, and the calculated $P^1{}_C(t)$ schedule were calculated. The load power values of the recommended $P_{LCR}(t)$ and the actual values of $P_{LC}(t)$, $P_{CB}(t)$, $P^1{}_C(t)$ are given in tabular form. The time intervals are given by the variables t_{12}, t_{23}, t_{34}, t_{45}, t_{56}, t_{67}, and t_{71}, which take on the value 1 at the corresponding time. The following auxiliary variables are also used:

$$q = \begin{cases} 1, \text{ if } Q^* \geq Q^*{}_d \\ 0, \text{ if } Q^* < Q^*{}_d \end{cases}, \ pv = \begin{cases} 1, \text{ if } P_{PVM}\eta_C \geq P^1{}_C \\ 0, \text{ if } P_{PVM}\eta_C < P^1{}_C \end{cases}, \ lc = \begin{cases} 1, \text{ if } P_C \geq P_{LC} \\ 0, \text{ if } P_C < P_{LC} \end{cases},$$

$$c = \begin{cases} 1, \text{ if } P_{LC} \geq P_{PVM}\eta_C \geq P^1{}_{C23} \\ 0, \text{ if } P_{LC} < P_{PVM}\eta_C < P^1{}_{C23} \end{cases}, \ s = \begin{cases} 1, \text{ if } P_{PVM}\eta_C > 0 \\ 0, \text{ if } P_{PVM}\eta_C \leq 0 \end{cases}, \ h = \begin{cases} 1, \text{ if } (P_{LC} - P_{LCR}) > 0 \\ 0, \text{ if } (P_{LC} - P_{LCR}) \leq 0 \end{cases},$$

$$qc = \begin{cases} 1, \text{ if } P_{LC} > P_C \\ 0, \text{ if } P_C \geq P_{LC} \end{cases}, \ w = \begin{cases} 1, \text{ if } W_{PV23} \cdot \eta_C / W_{CB23} \geq 1.5 \\ 0, \text{ if } W_{PV23} \cdot \eta_C / W_{CB23} < 1.5 \end{cases}$$

The variable w is precalculated. To measure the values of $Q^*{}_t$, $Q^*{}_5$, sample-and-hold schemes are used.

The current PV generation takes into account the following regulation:

$$P_{PV}\eta_C = P_{PVM} \cdot \eta_C \cdot (\bar{q} + qc) + (P_C + P_B)q \cdot lc,$$

where $P_B = U_B I_B$.

The value of added power is

$$P_C = P_{LC} \cdot t_{62} + P^1{}_C (t_{23} \cdot \overline{w} + t_{34} \cdot \overline{q} + t_{45} \cdot \overline{pv}) + P_{C56} \cdot t_{56} + (P^1{}_{C23} \cdot \overline{c} +$$
$$+ P_{LC} \cdot \overline{c} \cdot lc + P_{PVM} \cdot \eta_C \cdot c)w \cdot t_{23} + ((P_{PVM} \cdot \eta_C - P_B)q \cdot \overline{lc} + P_{LC} \cdot q \cdot lc)(t_{34} + t_{45}),$$

where $P_{C56} = I_{B56R} \cdot U_{BAV} + h(P_{LC} - P_{LCR})$.

The power consumed from the grid is $P_g = (P_{LC} - P_C)t_{26} + (P_{LC} + P_B)t_{62}$.

The current of the SB is

$$I_B = t_{62} \cdot \left(I_{B62R} \cdot \overline{s} + \frac{P_{PV}\eta_C}{U_B}s \right) + (\overline{q} + qc) \cdot t_{25} \cdot \frac{P_{PV}\eta_C - P_C}{U_B} + I_B(Q^*) \cdot q \cdot \overline{qc} \cdot t_{25} + t_{56}\frac{P_{C56}}{U_B}.$$

The SB model is constructed according to the principles set in [26,27]. Data sheets given by the manufacturer were used [31]: charge characteristics $I_{BC}(Q^*)$ and $U_{BC}(Q^*)$ at $I_B \geq 0$ and discharge characteristic $U_{BR}(Q^*)$ at $I_B < 0$, set in tabular form. The current can be calculated as

$$I_B = \begin{cases} I_B, & \text{if } Q^* < Q^*{}_d \\ I_B(Q^*), & \text{if } Q^* \geq Q^*{}_d \end{cases}.$$

The SB state of charge (*SOC*) taking into account energy losses is

$$Q = Q_0 + \int I^1{}_B dt,$$

where $I^1{}_B = I_B \cdot \eta_B$ if $I_B \geq 0$, and $I^1{}_B = I_B / \eta_B$ if $I_B < 0$.

To estimate the reduction in electricity costs at a single tariff rate (taking its value equal to 1), the $k_E = W_L / W_g$ coefficient was used ($W_L = \int_0^{24} P_{LC}dt$ is the energy consumed by the LO load per day (without taking into account the energy on the SB charge at night), and $W_g = \int_0^{24} P_g dt$ is the energy consumed by LO from the grid).

6. Simulation Results

To set the PV generation, archival data were used [30] with the selection of days when the generation by intervals was close to the average monthly generation (Table 1). These days correspond to the energy values without parentheses in Table 1. Values Q^*_6, k_E, and ρ for the case when $P_{LC} \leq 2P_{LIM}$ and the actual value of total load power $P_{LC} = P_{LCR}$ are given in Table 4.

Table 4. Simulation results.

Indicators	January	February	March	April	May	June	July	August	September	October	November	December
Possibilities of power increase P_{PVR} = 1 kW, W_B = 968 Wh												
ρ	1.7	1.7	1.8	1.9	1.95	1.95	1.95	1.95	1.9	1.75	1.7	1.7
k_E	1.105	1.4	1.636	2.128	2.618	2.684	2.767	2.291	1.918	1.586	1.096	1.149
Q^*_6, %	17.5	20	19.8	18.5	18.5	19.3	20	20	19.2	19.6	14	20
Constant value of power degree ρ = 1.6 (P_{PVR} = 0.86 kW, W_B = 800 Wh)												
k_E	1.098	1.354	1.608	2.316	2.821	2.804	2.946	2.321	2	1.523	1.088	1.133
Q^*_6, %	18.7	20.4	20.6	20	21	20	20	20	20.7	20	15	20

Oscillograms P_{LC}, P_{LCR}, P_C, P_{PVM}, P_{PV}, Q^*, I_B, and P_g (for clarity, P_g is shown as negative) are given as described below.

- In Figure 5 for the December day with a total generation twice below the average (W_{PV} = 500 Wh). In this case, k_E = 1.04, p = 1.5 (P_{PVR} = 0.86 kW, W_B = 800 Wh);

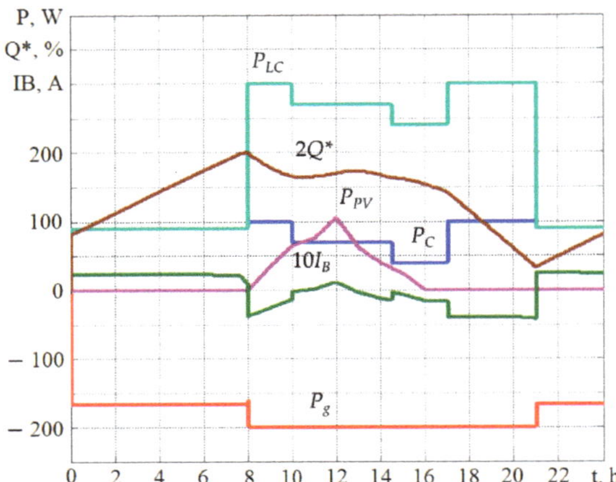

Figure 5. Oscillograms P_{LC}, P_C, P_{PV}, Q^*, I_B, and P_g for a December day with general generation $W_{PV} = 500$ Wh (Q^* and I_B values are shown at scale 2 and 10, respectively).

- In Figure 6 for the July day with a total generation in 3.3 times below the average ($W_{PV} = 1320$ Wh). In this case, $k_E = 1.22$, $p = 1.6$ ($P_{PVR} = 0.86$ kW, $W_B = 800$ Wh);

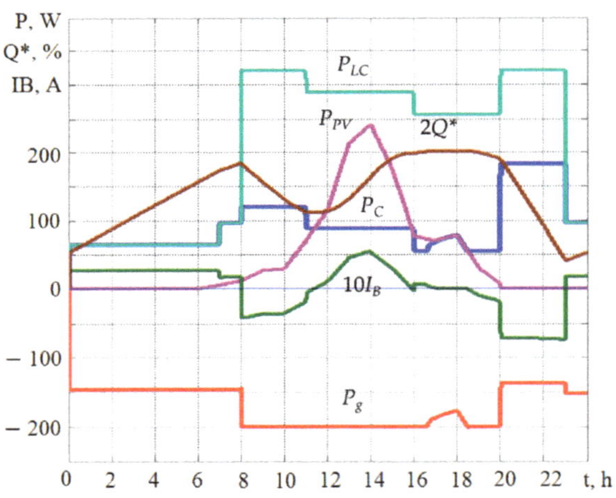

Figure 6. Oscillograms P_{LC}, P_C, P_{PV}, Q^*, I_B, and P_g for a July day with general generation $W_{PV} = 1320$ Wh (Q^* and I_B values are shown at scale 2 and 10, respectively).

- In Figure 7a for the May day with the generation, corresponding to the average monthly values at $P_{LC} = P_{LCR}$, $P_{PVR} = 1$ kW, $W_B = 968$ Wh, and at limit values $\rho = 1.95$ with $k_E = 2.618$;

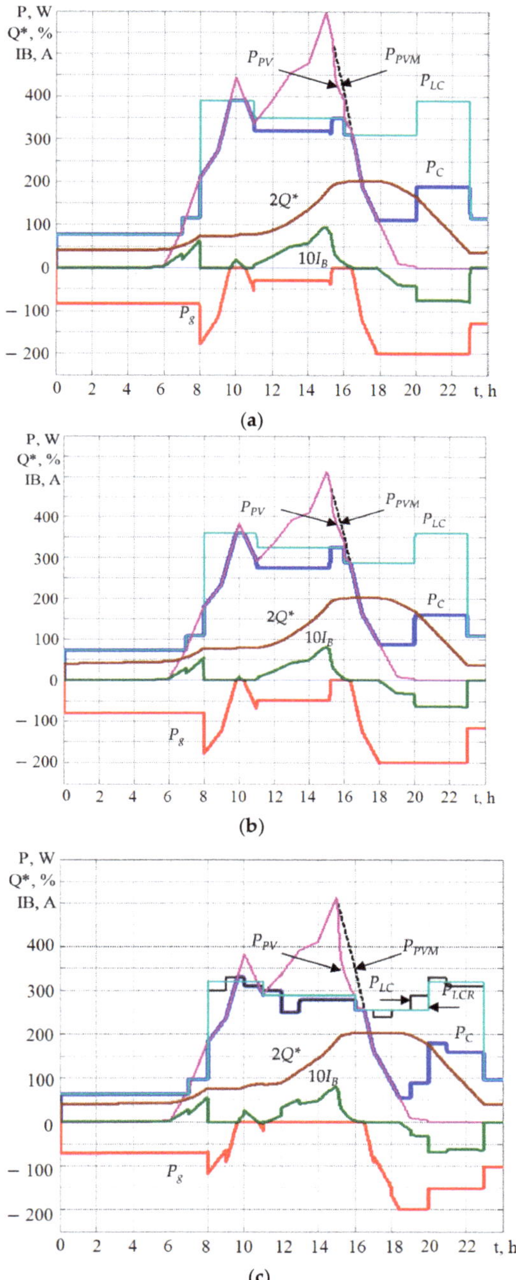

Figure 7. Oscillograms P_{LC}, P_{LCR}, P_C, P_{PVM}, P_{PV}, Q^*, I_B, and P_g for the May day with a generation corresponding to the monthly average (Q^* and I_B values are shown at scale 2 and 10, respectively): (**a**) $P_{LC} = P_{LCR}$, $P_{PVR} = 1$ kW, $W_B = 968$ Wh, and at limit values $\rho = 1.95$ with $k_E = 2.618$; (**b**) $P_{LC} = P_{LCR}$, $P_{PVR} = 0.86$ kW, $W_B = 800$ Wh, and at limit values $p = 1.8$ with $k_E = 2.356$; (**c**) $P_{LC} \neq P_{LCR}$, $P_{PVR} = 0.86$ kW, $W_B = 800$ Wh, and $p = 1.6$ with $k_E = 2.795$.

- In Figure 7b for the May day with the generation, corresponding to the average monthly values at $P_{LC} = P_{LCR}$, $P_{PVR} = 0.86$ kW, $W_B = 800$ Wh, and at limit values $p = 1.8$ with $k_E = 2.356$;
- In Figure 7c for the May day with the generation, corresponding to the average monthly values at $P_{LC} \neq P_{LCR}$, $P_{PVR} = 0.86$ kW, $W_B = 800$ Wh, and $p = 1.6$ with $k_E = 2.795$;
- In Figure 8 for the September day with the generation, corresponding to the average monthly values at $P_{LC} = P_{LCR}$, $P_{PVR} = 0.86$ kW, $W_B = 800$ Wh, and $\rho = 1.6$ with $k_E = 2$.

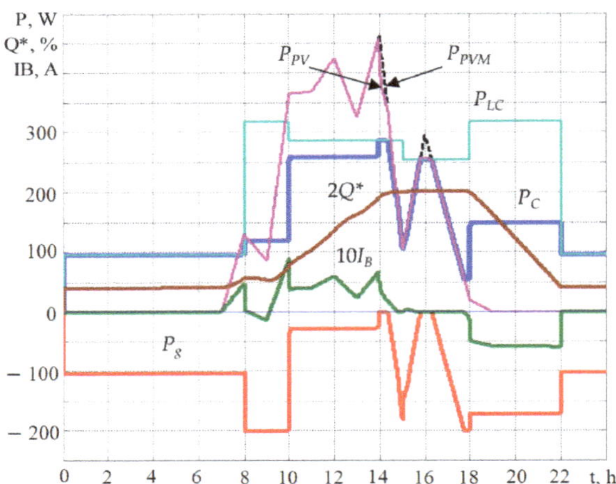

Figure 8. Oscillograms P_{LC}, P_{LCR}, P_C, P_{PVM}, P_{PV}, Q^*, I_B, and P_g for the September day with the generation, corresponding to the average monthly values at $P_{LC} = P_{LCR}$, $P_{PVR} = 0.86$ kW, $W_B = 800$ Wh, and $\rho = 1.6$ with $k_E = 2$ (Q^* and I_B values are shown at scale 2 and 10, respectively).

Changing the reference of the added power P_{C23} in the interval (t_2, t_3) from a fixed calculated value to the value corresponding to the actual PV schedule in the spring–autumn period allows increasing k_E by 2–4% with the exclusion of SB discharge.

7. Discussion of the Results of the Study on Increasing the Power of the LO Using PVS with SB

Increasing the load power of the LO above the limit of power consumption from the grid while reducing the installed power of PV and SB and decreasing the cost of paying for electricity, consumed by the LO, from the DG is possible due to the following:

- Use of the basic schedule of added power, tied to the PV generation. This reduces the energy required for its implementation. This allows increasing the degree of power increase while reducing the energy capacity of the SB;
- Limiting the degree of power increase at an intermediate value with a decrease in the installed power of the PV and SB. This provides an improvement in the use of PV energy without increasing the cost of electricity, consumed from the grid;
- Referencing the current value of the added load power and the *SOC* value of the battery at time intervals, taking into account the predicted and actual PV generation. Due to the change in certain time intervals of the added power reference from the calculated value to a value that repeats the law of the change in the PV generation power, this contributes to more complete use of the PV energy;
- Exclusion of the SB charge at night with the average monthly PV generation in the spring–summer–autumn period helps to reduce electricity consumption from the grid. This excludes battery discharge during the hours of the morning load peak, which helps to reduce the number of battery discharge cycles and increase its service life.

This article is a development of previous studies [19,26], which considered increasing the efficiency of hybrid PVS with SB for the needs of local facilities. This was achieved by reducing the cost of electricity consumed from the grid when using the PV generation forecast. A common problem is the significant overestimation of the PV power in relation to the load power of the LO, which is necessary for use in conditions of low PV generation.

As a result, even with medium PV generation, there is a significant underutilization of PV energy with the need for regulation of PV generation. A feature of the proposed solutions is a change in the general approach to the use of PV and SB energy when the load power, added to the limit, is formed. Directed formation of the graph of added power allows reducing the installed power of PV and SB.

There are certain limitations regarding the use of the results of the work, as described below.

- An object was considered with the main load in the daytime in the presence of peak loads in the morning and evening hours. At the same time, it is possible to charge the SB at night within the limit on consumption from the grid;
- The possibilities of increasing power are seasonal in nature with a maximum value in the period spring–summer–autumn;
- The optimal choice of values for the degree of increase in power (ρ) and the installed PV power (m_P) involves taking into account many factors, including the costs of acquisition and ongoing maintenance. Such a task was not set in this work, and the approach was simplified for evaluation;
- The assessment of a possible reduction in the cost of paying for electricity during the year was somewhat simplified and was performed for one tariff rate. We considered the days when the PV generation corresponds to the average monthly values for the accepted time intervals;
- The implementation of the control system assumes an "open" structure of the relevant channels of control;
- When modeling, it was assumed that the graph of the power generated by the PV corresponds to the forecast and does not change during the day.

Further development of this work is connected with the optimization of system parameters. It is also required to study the possibilities of correction of deviations in the values of the actual PV generation according to the forecast data and possible changes in the current forecast during the day.

8. Conclusions

With the selected parameters of the PVS, it is possible to increase the power of the LO over the limit for consumption from the grid up to 1.7–1.95 times. This depends on the average monthly PV generation of during the year. Limiting the degree of power increase to a value of 1.6, when choosing the parameters, allows reducing the installed power of the PV and SB. The possibility of increasing the degree of power increase to a value of 1.8 (if necessary) remains in this case.

It is advisable to select the parameters of the PVS on the basis of the data on the average monthly PV generation for the taken schedule of the load. It is possible to use archival data from web resources with open access at any point location of the object. The ratio of PV power and added load power is determined on the basis of PV generation in the transitional seasons (in this case, March and October). The energy capacity of the SB is determined by the power of the added load from the condition of the sufficiency of its operation in the pre-evening hours and evening peak hours with a given limitation of the DOD battery. The use of the basic schedule of the added power with reference to solar activity, while maintaining the resulting power of the LO, makes it possible to reduce the energy capacity of the SB. Therefore, when setting the degree of power increase $\rho = 1.7$ for winter, the energy capacity can be reduced by 13.5%. Limiting the value of ρ at the level of $\rho = 1.6$ allows reducing the installed power of the PV and the SB. When converted to the same total load power, the reduction for P_{PVR} is 9.4%, and that for W_{BR} is 14%.

The principle for controlling the power consumed by the LO from the grid in accordance with the reference value of the added load power and the total load power of the LO was developed. The reference of the value of the added power can be determined from the condition for the formation of the $SOC(t)$ graph of the SB according to the PV generation forecast data.

At the same time, at certain time intervals, reference of the value of the added power can be carried out according to the schedule of the PV generation. This contributes to the maximum use of its energy. With the average monthly PV generation in the spring–autumn period, the SB discharge during the hours of the morning load peak is not used. Accordingly, the nighttime SB charge from the grid is not used. This helps to reduce consumption from the grid and reduce the number of deep discharge cycles, which contributes to longer battery life.

On the basis of a formalized description of energy processes in steady-state conditions for the daily cycle of the system's functioning, the simulation of the system was completed. The simulation results showed that choosing the PVS parameters on the basis of $\rho = 1.7$ by setting $\rho = 1.7$–1.95 depending on the average monthly PV generation reduced the cost of electricity, consumed from the grid, for one rate of payment by 1.1 to 2.68 times. When choosing the PVS parameters on the basis of $\rho = 1.6$ and a constant degree of power increase during the year, it is possible to reduce electricity costs by up to 7% in the summer.

Author Contributions: Conceptualization, O.S., I.S. and J.G.; methodology, O.S., I.S. and J.G.; validation, O.S. and I.S.; formal analysis, I.S. and K.K.; investigation, O.S., I.S., K.K. and F.P.; resources, I.S. and K.K.; data curation, O.S. and I.S.; writing—original draft preparation, O.S., I.S. and F.P.; writing—review and editing, J.G.; visualization, I.S. and K.K.; supervision, J.G. All authors have read and agreed to the published version of the manuscript.

Funding: This publication was issued thanks to support from the Cultural and Educational Grant Agency of the Ministry of Education of the Slovak Republic through the projects "Implementation of modern methods of computer and experimental analysis of properties of vehicle components in the education of future vehicle designers" (Project No. KEGA 036ŽU-4/2021) and "Development of advanced virtual models for studying and investigation of transport means operation characteristics" (Project No. KEGA 023ŽU-4/2020). This research was also supported by the Slovak Research and Development Agency of the Ministry of Education, Science, Research, and Sport of the Slovak Republic in Educational Grant Agency of the Ministry of Education of the Slovak Republic through the project "Investigation of the properties of railway brake components in simulated operating conditions on a flywheel brake stand" (Project No. VEGA 1/0513/22).

Institutional Review Board Statement: Not applicable.

Informed Consent Statement: Not applicable.

Data Availability Statement: Not applicable.

Acknowledgments: This work was supported by the Ministry of Education and Science of Ukraine in the joint Ukrainian–Slovak R&D project "Energy management improvement of hybrid photovoltaic systems of local objects with storage batteries" (0122U002588).

Conflicts of Interest: The authors declare no conflict of interest.

References

1. Rao, B.H.; Selvan, M.P. Prosumer Participation in a Transactive Energy Marketplace: A Game-Theoretic Approach. In Proceedings of the IEEE International Power and Renewable Energy Conference, Karunagappally, India, 30 October–1 November 2020.
2. Khezri, R.; Mahmoudi, A.; Aki, H. Optimal planning of solar photovoltaic and battery storage systems for grid-connected residential sector: Review, challenges and new perspectives. *Renew. Sustain. Energy Rev.* **2022**, *153*, 111763. [CrossRef]
3. ABB Solar Inverters. Product Manual REACT-3.6/4.6-TL (from 3.6 to 4.6 kW). Available online: https://www.abb.com/solarinverters (accessed on 9 March 2022).
4. Conext, S.W. Hybrid Inverter. Available online: https://www.se.com/ww/en/product-range-presentation/61645-conext-sw/ (accessed on 9 March 2022).
5. Zeng, Z.; Yang, H.; Zhao, R.; Cheng, C. Topologies and control strategies of multi-functional grid-connected inverters for power quality enhancement: A comprehensive review. *Renew. Sustain. Energy Rev.* **2013**, *24*, 223–270. [CrossRef]

6. Ma, T.-T. Power quality enhancement in micro-grids using multifunctional DG inverters. *Lect. Notes Eng. Comput. Sci.* **2012**, *2196*, 996–1001.

7. Shavelkin, A.; Jasim, J.M.J.; Shvedchykova, I. Improvement of the current control loop of the single-phase multifunctional grid-tied inverter of photovoltaic system. *East.-Eur. J. Enterp. Technol.* **2019**, *6*, 14–22. [CrossRef]

8. Belaidi, R.; Haddouche, A. A multi-function grid-connected PV system based on fuzzy logic controller for power quality improvement. *Przeg. Elektr.* **2017**, *93*, 118–122. [CrossRef]

9. Vigneysh, T.; Kumarappan, N. Grid interconnection of renewable energy sources using multifunctional grid-interactive converters: A fuzzy logic based approach. *Electr. Power Syst. Res.* **2017**, *151*, 359–368. [CrossRef]

10. Shavolkin, O.; Shvedchykova, I. Improvement of the Three-Phase Multifunctional Converter of the Photoelectric System with a Storage Battery for a Local Object with Connection to a Grid. In Proceedings of the 25th IEEE International Conference on Problems of Automated Electric Drive. Theory and Practice, PAEP 2020, Kremenchuk, Ukraine, 21–25 September 2020; pp. 1–6.

11. Shavolkin, O.; Shvedchykova, I.; Demishonkov, Y. Energy Management of Hybrid Three-Phase Photoelectric System with Storage Battery to Meet the Needs of Local Object. In Proceedings of the 20th IEEE International Conference on Modern Electrical and Energy Systems, MEES 2021, Kremenchuk, Ukraine, 21–24 September 2021.

12. Lliuyacc, R.; Mauricio, J.M.; Gomez-Exposito, A.; Savaghebi, M.; Guerrero, J.M. Grid-forming VSC control in four-wire systems with unbalanced nonlinear loads. *Electr. Power Syst. Res.* **2017**, *152*, 249–256. [CrossRef]

13. Guerrero-Martinez, M.A.; Milanes-Montero, M.I.; Barrero-Gonzalez, F.; Miñambres-Marcos, V.M.; Romero-Cadaval, E.; Gonzalez-Romera, E. A smart power electronic multiconverter for the residential sector. *Sensors* **2017**, *17*, 1217. [CrossRef]

14. Luthander, R.; Widén, J.; Nilsson, D.; Palm, J. Photovoltaic self-consumption in buildings: A review. *Appl. Energy* **2015**, *142*, 80–94. [CrossRef]

15. Yang, X.; Jiang, F.; Liu, H. Short-term power prediction of photovoltaic plant based on SVM with similar data and wavelet analysis. *Prz. Elektrotech.* **2013**, *89*, 81–85.

16. Mellit, A.; Pavan, A.M.; Lughi, V. Deep learning neural networks for short-term photovoltaic power forecasting. *Renew. Energy* **2021**, *172*, 276–288. [CrossRef]

17. Zsiborács, H.; Pintér, G.; Vincze, A.; Birkner, Z.; Baranyai, N.H. Grid balancing challenges illustrated by two European examples: Interactions of electric grids, photovoltaic power generation, energy storage and power generation forecasting. *Energy Rep.* **2021**, *7*, 3805–3818. [CrossRef]

18. Michaelson, D.; Mahmood, H.; Jiang, J. A Predictive Energy Management System Using Pre-Emptive Load Shedding for Islanded Photovoltaic Microgrids. *IEEE Trans. Ind. Electron.* **2017**, *64*, 5440–5448. [CrossRef]

19. Shavolkin, O.; Shvedchykova, I.; Jasim, J.M.J. Improved control of energy consumption by a photovoltaic system equipped with a storage device to meet the needs of a local facility. *East.-Eur. J. Enterp. Technol.* **2021**, *2*, 6–15. [CrossRef]

20. Forecast. Solar. Available online: https://forecast.solar/ (accessed on 9 March 2022).

21. Iyengar, S.; Sharma, N.; Irwin, D.; Shenoy, P.; Ramamritham, K. SolarCast—An open web service for predicting solar power generation in smart homes. In Proceedings of the 1st ACM Conference on Embedded Systems for Energy-Efficient Buildings, BuildSys 2014, Memphis, TN, USA, 3–6 November 2014; pp. 174–175.

22. Nicolson, M.L.; Fell, M.J.; Huebner, G.M. Consumer demand for time of use electricity tariffs: A systematized review of the empirical evidence. *Renew. Sustain. Energy Rev.* **2018**, *97*, 276–289. [CrossRef]

23. Davis, M.J.M.; Hiralal, P. Batteries as a Service: A New Look at Electricity Peak Demand Management for Houses in the UK. *Procedia Eng.* **2016**, *145*, 1448–1455. [CrossRef]

24. Lorenzi, G.; Silva, C.A.S. Comparing demand response and battery storage to optimize self-consumption in PV systems. *Appl. Energy* **2016**, *180*, 524–535. [CrossRef]

25. Badawy, M.O.; Cingoz, F.; Sozer, Y. Battery storage sizing for a grid tied PV system based on operating cost minimization. In Proceedings of the 2016 IEEE Energy Conversion Congress and Exposition, Milwaukee, WI, USA, 18–22 September 2016.

26. Shavelkin, A.A.; Gerlici, J.; Shvedchykova, I.O.; Kravchenko, K.; Kruhliak, H.V. Management of power consumption in a photovoltaic system with a storage battery connected to the network with multi-zone electricity pricing to supply the local facility own needs. *Electr. Eng. Electromech.* **2021**, *2*, 36–42. [CrossRef]

27. Shavolkin, O.; Shvedchykova, I.; Demishonkova, S.; Pavlenko, V. Increasing the efficiency of hybrid photoelectric system equipped with a storage battery to meet the needs of local object with generation of electricity into grid. *Przeg. Elektr.* **2021**, *97*, 144–149. [CrossRef]

28. Traore, A.; Taylor, A.; Zohdy, M.; Peng, F. Modeling and Simulation of a Hybrid Energy Storage System for Residential Grid-Tied Solar Microgrid Systems. *J. Power Energy Eng.* **2017**, *5*, 28–39. [CrossRef]

29. Miñambres-Marcos, V.M.; Guerrero-Martínez, M.Á.; Barrero-González, F.; Milanés-Montero, M.I. A grid connected photovoltaic inverter with battery-supercapacitor hybrid energy storage. *Sensors* **2017**, *17*, 1856. [CrossRef] [PubMed]

30. Photovoltaic Geographical Information System. Available online: https://re.jrc.ec.europa.eu/pvg_tools/en/tools.html#SA (accessed on 9 March 2022).

31. Data Sheet. Lithium Iron Phosphate (LiFePo4) Battery 12.8 V 150 Ah. Available online: https://www.enix-energies.com (accessed on 9 March 2022).

Article

Acoustic Signature and Impact of High-Speed Railway Vehicles in the Vicinity of Transport Routes

Krzysztof Polak [1],* and Jarosław Korzeb [2]

[1] Railway Track and Operation Department, Railway Institute, 04-275 Warsaw, Poland
[2] Faculty of Transport, Warsaw University of Technology, 00-662 Warsaw, Poland; jaroslaw.korzeb@pw.edu.pl
* Correspondence: kpolak@ikolej.pl

Abstract: In this paper, an attempt is undertaken to identify the acoustic signature of railway vehicles travelling at 200 km/h. In the framework of conducted experimental research, test fields were determined, measurement apparatus was selected and a methodology for making measurements was specified, including the assessment of noise emission on curved and straight track for electric multiple units of Alstom type ETR610-series ED250, the so-called Pendolino. The measurements were made with the use of an acoustic camera and a 4×2 microphone array, including four equipped measurement points and two microphones located at the level of the head of the rail and at a height of 4 m above this level. As a result of the conducted experimental research, the dominant noise sources were identified and amplitude–frequency characteristics for these sources were determined by dividing the spectrum into one-third octave bands in the range from 20 Hz to 20 kHz. The paper also considers issues related to the verification of selected models of noise assessment in terms of their most accurate reflection of the phenomenon of propagation in close surroundings. On the basis of conducted experimental studies, the behaviour of selected models describing the change of sound level with frequency division into one-third octave bands as a function of variable distance of observer from the railway line on which high-speed railway vehicles are operated was verified. In addition, the author's propagation model is presented together with a database built within the scope of the study, containing the actual waveforms in the time and frequency domain.

Keywords: railway noise; high-speed railways; environmental impact

Citation: Polak, K.; Korzeb, J. Acoustic Signature and Impact of High-Speed Railway Vehicles in the Vicinity of Transport Routes. *Energies* **2022**, *15*, 3244. https://doi.org/10.3390/en15093244

Academic Editors: Larysa Neduzha and Jan Kalivoda

Received: 9 March 2022
Accepted: 25 April 2022
Published: 28 April 2022

Publisher's Note: MDPI stays neutral with regard to jurisdictional claims in published maps and institutional affiliations.

Copyright: © 2022 by the authors. Licensee MDPI, Basel, Switzerland. This article is an open access article distributed under the terms and conditions of the Creative Commons Attribution (CC BY) license (https://creativecommons.org/licenses/by/4.0/).

1. Introduction

According to the European Environment Agency report [1] results, railway transport is the second-largest source of noise in Europe. It affects about 4% of the population during the day–evening–night period and 3% during the night. Almost 22 million people are exposed to noise intensity of at least 55 dB during the day–evening–night period and about 16 million people to noise at night (of intensity of minimum 50 dB). The scale of this phenomenon is influenced by a number of factors, including the technical condition of the railway superstructure and rolling stock, landforms and land use, and the high speed of railway vehicles [2,3].

The share of high-speed railway vehicles in the rolling stock of operators providing public transport services is continuously increasing. At present, in Poland, on railway lines with the highest speed, there may operate Alstom type ETR610 series ED250 vehicles (the so-called Pendolino). These vehicles can travel at a maximum speed of 250 km/h; however, due to technical limitations of the railway infrastructure, their maximum operating speed on the Central Railway Line (railway line no. 4 Grodzisk Mazowiecki-Zawiercie) is equal to 200 km/h [4].

The literature analysis showed that there are many works and studies aimed at identifying or determining the acoustic signature of railway vehicles [5–12]. The authors of this work carried out tests in accordance with the methodology specified in the ISO 3095 standard [13]. According to the standard [13], measurements should be made at one or two measurement points, with one microphone in each. The measurement points should be located at a distance of 7.5 m from the track axis at a height of 1.2 m from the rail head and/or at a distance of 25 m at a height of 3.5 m from the rail head.

These tests mainly involved high-speed vehicles travelling at speeds exceeding 200 km/h [11,14–16]. In addition, a review of literature publications and studies conducted so far in Poland reveals a lack of available studies on the characteristics of the dominant noise sources generated by Alstom ETR610 series ED250 high-speed railway vehicles. In this study, measurements were made not only to examine the noise level of the train passage, but an attempt was made to identify the acoustic signature as accurately as possible, hence a nonstandard location of the measurement points was proposed.

External rail noise may arise from a number of sources, including rolling noise, impact noise, traction noise, aerodynamic noise or braking and starting noise. The noise of a particular unit depends, among other things, on the design of the unit and the operating conditions, and therefore a characterisation of the noise emission of the unit is an important element in minimising the acoustic impact [13].

In this article, selected results and conclusions were discussed from experimental research conducted with the use of an acoustic camera and a 4×2 microphone array (four measurement points at a distance of 5 m, 10 m, 20 m and 40 m from the track axis, two microphones located at the height of the rail head and at a height of 4 m from the rail head). The aim of this paper is to characterise the acoustic signaling of high-speed railway vehicles travelling at 200 km/h in Polish conditions. The main sources of noise generation were identified and their spectrum described by amplitude–frequency characteristics. Such an approach is nonstandard due to the tertial resolution of the identified propagation characteristics. Acoustic signals coming from the studied object were measured, for which no complex work has been carried out so far to assess their acoustic impact on the surroundings. Recognition of the acoustic signature of high-speed railway vehicles will allow for the building of a model of the source of acoustic impacts typical for the analysed vehicles. Due to the very good condition of the track in the studied area and commissioning of the first section admitted to traffic at high speed, it was possible to build a reference source of impact for the purposes of noise modelling as well as to establish the reference threshold against which passes of trains operating with a higher degree of wheel wear or on tracks with a higher degree of rail wear can be evaluated in the future.

Such information can be used to establish the relationship between environmental impacts and the technical condition of the rolling stock in operation. Complementing the knowledge on their acoustic impact on the surroundings will allow more effective work on minimising the negative impact of noise generated by this type of vehicles.

The experimental investigations carried out and the database built within the framework of the study, containing real waveforms in the time and frequency domain, served as a source of information on the generated impacts for the purposes of modelling noise propagation in close surroundings. Selected models of railway noise evaluation were analysed in terms of their most accurate reflection of the propagation phenomenon. On the basis of conducted experimental studies, the behaviour of selected models describing the change of sound level in the i-th frequency band as a function of variable distance of the observer from the railway line on which high-speed railway vehicles are operated was verified.

Additionally, on the basis of the obtained measurement results, an own model of noise propagation in one-third octave bands within the audible band (20 Hz–20 kHz) was built, which was characterised by the best accuracy of reflecting the phenomenon of noise propagation in the range of measurement results in particular measurement points and beyond them.

2. Materials and Methods

2.1. Technical Specification of the Research Objects

The objects of the experimental research were Alstom's railway vehicles type ETR610-series ED250-Pendolino (Figure 1). These vehicles are managed by the largest operator in Poland, PKP Intercity S.A. These are bidirectional electric multiple units consisting of seven sections, including four engine sections (two end sections on both sides) and three trailer sections [17,18].

Figure 1. Object of the research: an Alstom vehicle ETR610-series ED250-Pendolino [photo: own elaboration].

Normal-gauge vehicles (suitable for operation on tracks with a gauge of 1435 mm) are equipped with eight asynchronous, three-phase traction motors with a power of 708 kW each. The length of the vehicle is 187.4 m and its weight is 395.5 tones [19]. The vehicles use two-axle bogies with one driving axle and one rolling axle, with a very low mass value and equipped with torsion dampers, which ensure appropriate dynamic characteristics of the vehicle on curves at speeds of 250 km/h [17,20,21].

There were two sections of a railway track selected for the purpose of the experimental research (straight section and curved section), located on the railway line no. 4 Grodzisk Mazowiecki–Zawiercie (Central Trunk Line–CMK), section Grodzisk Mazowiecki–Idzikowice, which is an electrified and mostly double-track line, with maximum design speed of 250 km/h (Figure 2) [22]. Details of the study sites are described, among others, in [2].

Figure 2. Railway track on the 18 + 600 km area of experimental research (a curve).

2.2. Methodology of the Conducted Research Area

The measurements of acoustic signals were conducted on railway line no. 4 Grodzisk Mazowiecki–Zawiercie, section Grodzisk Mazowiecki–Idzikowice, in two locations:

- curved section (R = 4000 m)—approx. 18 + 600 km (loc. Świnice, Długa str.);
- straight section—approx. 21 + 300 km (loc. Szeligi, Dojazdowa str.).

The registration of acoustic events from high-speed railway vehicles was made with the use of measurement apparatus, consisting of the following devices:

- acoustic camera Noise Inspector Bionic M-112;
- two Svantek sound-level meters SVAN 979;
- Svantek four-channel sound-level meter SVAN 958A;
- two Svantek sound-level meters SVAN 955;
- measurement microphones (electroacoustic transducers) from Svantek—8 pieces;
- Svantek 1st class acoustic calibrator SV 36;
- speedometer;
- weather station from Davis Instruments Vantage Pro2.

Detailed information concerning e.g., location of the testing ground, weather conditions, is described in the publications [2,3].

During noise measurements with a microphone array (4 × 2), eight measurement microphones operating in the audible band were used. The microphones were placed at a height of 4 m (from the rail head) and at the rail-head height (approx. 0.8 m from the ground). Additionally, in order to locate the main sources of noise, an acoustic camera located about 20 m from the track axis was used. A schematic of the measurement cross-section during the conducted experimental research is shown in Figure 3. The results obtained were subject to acquisition using Svantek's dedicated software, SvanPC++, which enabled the generation of data for further analysis and processing, as intended. The measurement of noise with a microphone array was performed using the continuous method, i.e., all acoustic phenomena were measured during the experimental studies of each run, while the frequency spectrum was analysed in one-third octave bands for each second. Registration of individual journeys (type of vehicle, time) made it possible, in the further part of the work, to select only the passages of the tested object. A schematic of the measurement path using the microphone array is shown in the diagram (Figure 4).

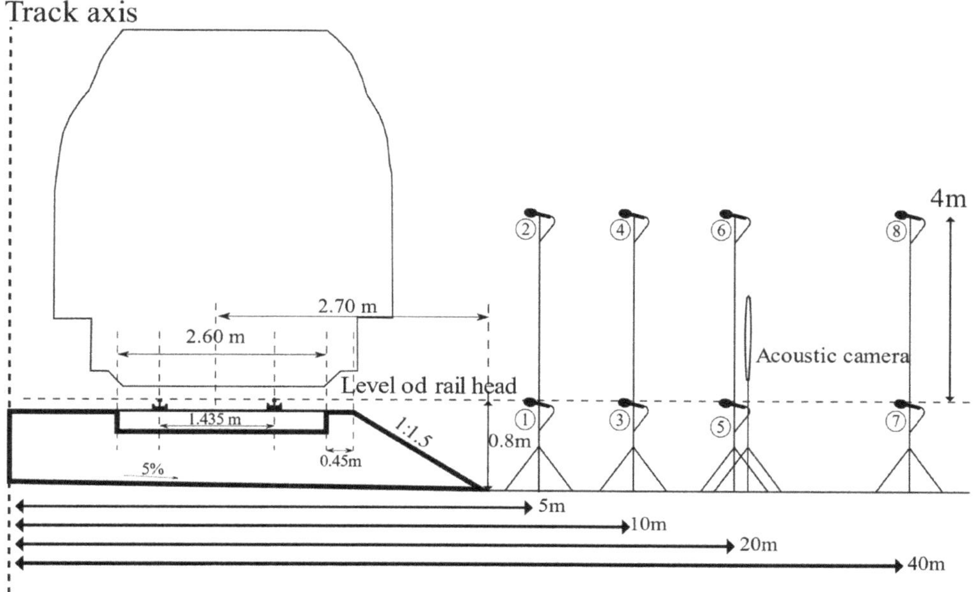

Figure 3. Instrumented cross section during experimental testing [Source: own elaboration].

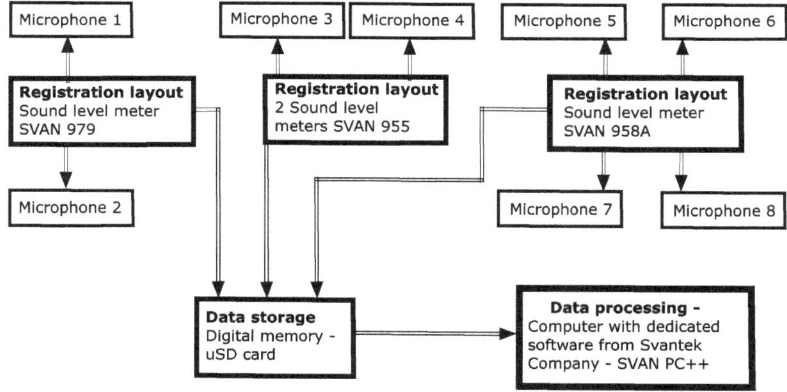

Figure 4. Scheme of the measurement circuit carried out with the sound-level meters [source: own elaboration].

3. Results and Discussion

As a result of the experimental study, the acoustic signature of high-speed vehicles operating in Polish conditions was identified by indicating the main sources of noise generated by the test object using an acoustic camera and obtaining a time history recorded for each pass of the tested vehicle separately (with a step of 1 s), containing the equivalent continuous sound level L_{Aeq} for each second of the pass and obtaining the frequency spectrum of noise for each second, divided into one-third octave bands within the audible band, i.e., from 20 Hz to 20 kHz.

3.1. Identification of Main Noise Sources

A Noise Inspector Bionic M-112 acoustic camera equipped with 112 directional microphones was used to determine the main noise sources. Thanks to the mounted optical camera, the image is assigned to the distribution of acoustic pressure levels. The acoustic camera uses, among others, the beamforming method for measurements, which allows results to be obtained in the frequency range of 400 Hz to 24 kHz, with a sampling rate of approximately 12 Hz.

The beamforming method consists in processing a spatial-temporal signal recorded by a microphone array. Each microphone of the acoustic array, having a specific position relative to its centre, is assigned a time delay which allows it to focus the signal beam in the direction of acoustic wave propagation [5,19,23–27]. By using a time delay, the measured signal is the same for all microphones used. The signal received by the microphones, delayed by an appropriate amount of time, is summed at the output of the array and thus amplified compared to a measurement with only one microphone [25,27–29].

On the basis of the target measurements carried out in August 2019, the main source of noise generated by high-speed railway vehicles travelling at around 200 km/h was defined. In order to better illustrate the results obtained, the acoustic events [dBA] characterised by the highest noise levels were grouped into three frequency ranges, i.e., 500–1000 Hz, 1000–2000 Hz and 2000–3000 Hz (Figure 5).

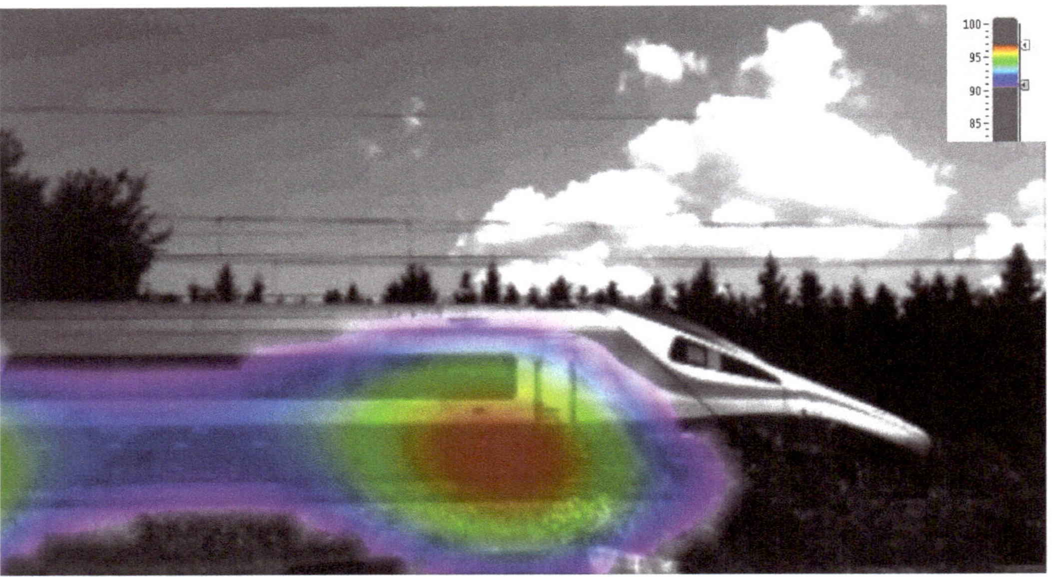

Figure 5. Sound-level distribution in the frequency range 500–1000 Hz—straight section [dBA].

On the basis of experimental tests carried out with the use of an acoustic camera, the main sources of noise [dBA] coming from high-speed vehicles moving on a straight section and along a curve at a speed of 200 km/h were identified. As a result of the experimental studies, it was indicated that the noise resulting from contact phenomena at the wheel–rail interface (rolling noise) and from the operation of drive units are the dominant sources of sound in the frequency range of 500–3000 Hz.

3.2. Noise Propagation in the Immediate Vicinity

In the framework of conducted experimental research, 20 passes of high-speed railway vehicles were obtained for a straight section and a curve, from which the 10 most authoritative results were selected, for which eight signals (acoustic pressure levels–[dB]), 20 s long, were obtained in 31 one-third octave bands from 20 Hz to 20 kHz in two measurement cross sections (straight section and curve). In addition, the equivalent continuous sound level [dBA] was measured at all points in the measurement cross section. The obtained data were subjected to further detailed analysis.

As a first step, the obtained data were verified in order to extract representative acoustic signals [dB] generated by high-speed railway vehicles from the recorded 20-s passes. According to relation (1), the total time of passage of the tested object through the measurement cross section (t) was determined, which amounted to about 3.4 s.

$$t = \frac{l}{v} \tag{1}$$

A representative sample of acoustic signals from speeding vehicles was assumed to be 4 s, which was also confirmed by the results obtained (Figure 6).

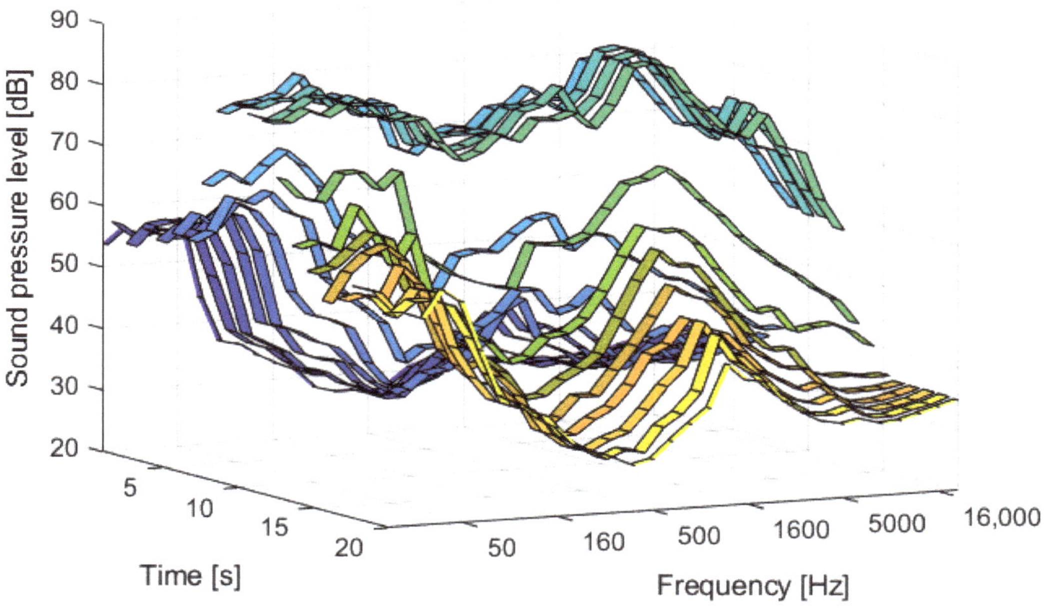

Figure 6. Amplitude–frequency distribution of acoustic pressure levels for a 20-s passage on the straight section.

The analysis made it possible to obtain 4-s passes representative of high-speed railway vehicles travelling on a straight section and a curve. On the diagram below (Figure 7) the distribution of acoustic pressure levels for a selected representative passage (4 s) is shown.

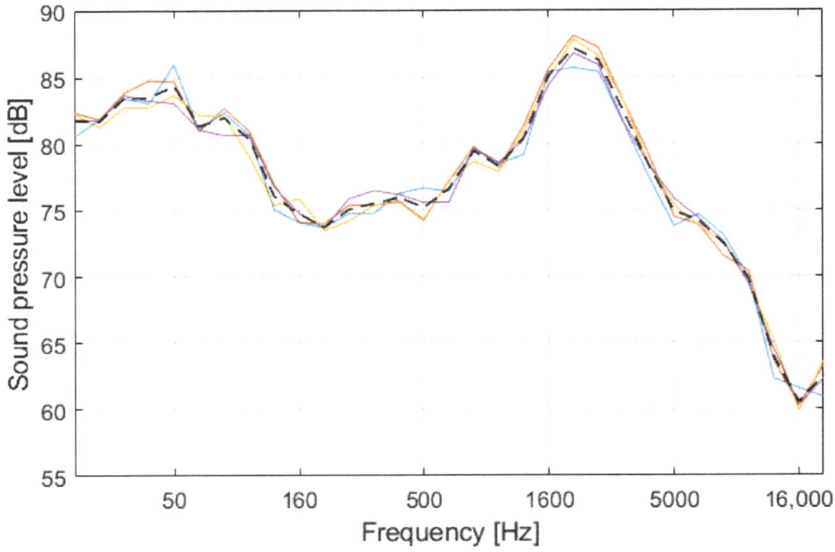

Figure 7. Distribution of acoustic pressure levels for 4 s representative passages on the straight section.

The standard deviation of the sample was then determined on the basis of the obtained data from the representative samples, which show the dispersion of the acoustic pressure levels for the analysed high-speed railway vehicle crossings on the straight section and the curve depending on the tertiary band. The in-sample standard deviation (2), which is the square root of the in-sample variance, was determined according to the formula [30–32]:

$$s = \sqrt{s^2} = \sqrt{\frac{\sum_{i=1}^{N}\left(\overline{x}_i - x\right)^2}{n - 1}} \tag{2}$$

An example of the distribution of the standard deviation for a selected journey is shown in Figure 8.

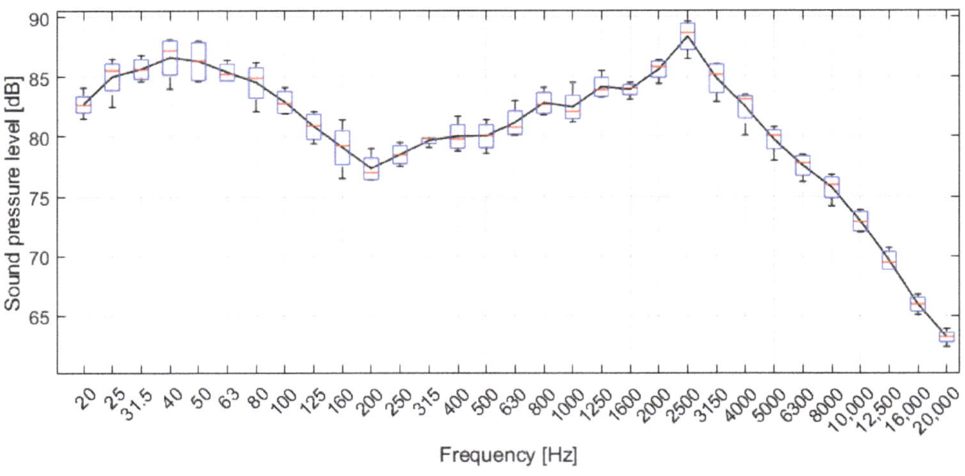

Figure 8. Distribution of standard deviation for a selected straight-section passage.

The analysis of obtained standard-deviation data showed that differences in acoustic pressure levels in given one-third octave bands in most cases did not exceed 2 dB—97.4%; only 2.6% of data indicated exceedances above 2 dB. Exceedances above 2 dB concerned 13 cases out of 496, which were within the range from 2.1 dB to 2.7 dB.

Taking into account that the values of perceived changes in acoustic pressure levels by the human ear vary from 2 dB [31] to 3 dB [33] (depending on the predisposition of the recipient), it was possible to consider the results as reliable.

In order to verify which one-third octave bands' frequencies have the highest acoustic pressure levels, the average spectra of the sound pressure levels were determined for the tested railway vehicle at the level of the head of the rail and at a height of 4 m from the level of the head of the rail, which moved along a straight section and a curve. The average acoustic pressure levels in one-third octave bands at rail-head height for a straight section are shown in Table 1 and Figure 9.

Table 1. Average acoustic pressure levels [dB] in one-third octave bands at rail-head height for a straight section.

One-Third Octave Bands [Hz]	5	10	20	40
20	99.8	95.4	94	89.7
25	99.8	96.6	94.3	90.9
31.5	100.6	97.1	95.1	91
40	101.4	98.3	95.9	92
50	102	99.1	96.3	92.6
63	100.5	100	94.9	91.2
80	100	98.5	94.3	89.9
100	98.3	97.9	93.6	89.6
125	96	95.1	91.4	85.4
160	93.2	92.5	90.7	83.7
200	91.6	90.5	88.3	81.4
250	93.3	89.4	85.8	79.2
315	94.1	88.4	83.3	76.7
400	94.2	87.3	82.5	75.7
500	95	88	83.5	77.3
630	95.9	90.2	85.2	77.6
800	97.9	93.9	87.4	79.8
1000	96.5	94.5	87.9	80.5
1250	97.4	95.2	88.8	81.2
1600	100.3	98.5	92.7	86.9
2000	101.6	99.7	95.4	88.1
2500	103	100.1	93.8	87.1
3150	99.8	97.3	89	86.3
4000	96.6	93.4	84.5	80.4
5000	93.8	90.1	80.7	77
6300	92	87.7	77.6	72.6
8000	91.6	87.4	76.5	68.9
10,000	88.4	84.4	71.5	65.1
12,500	84.8	80	66.4	58.7
16,000	80.7	75.6	57.5	55.2
20,000	78.1	70.2	54.6	58.3

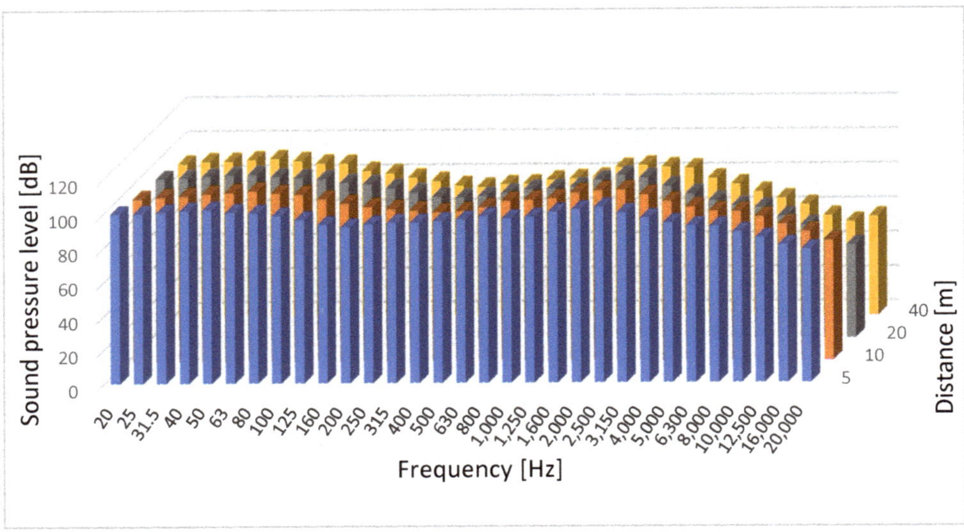

Figure 9. Average acoustic pressure levels in one-third octave bands at rail-head height for a straight section.

For both situations studied (straight section and curve), at the level of the rail head in the close vicinity of the railway line (at a distance of 20–40 m from the track centre), the low bands between 20 Hz and 200 Hz are dominant, with levels ranging from 88.3 dB to 96.9 dB. In the case of medium-frequency bands for the straight section and curve, at a distance of 20–40 m, high values of acoustic pressure levels were recorded for frequencies from 1600 Hz to 3150 Hz (89.0 dB–96.9 dB). In addition, for a railway line located on a curve, high acoustic pressure levels were recorded for frequencies between 630 Hz and 1250 Hz (89.6 dB–95.0 dB).

For measurement points located 4 m from rail-head level (at a distance of 20–40 m), the highest acoustic pressure levels are seen in the low bands from 20 Hz to 100 Hz and the medium bands, i.e., from 500 Hz to 3150 Hz.

In addition, the research carried out has shown that for the passage of high-speed vehicles along a curve with a radius of approximately R = 4000 m, the dominant frequencies characterising the passage of a high-speed railway vehicle can be identified.

Based on the analysis of the differences in acoustic pressure levels (difference of more than 3 dB) for individual one-third octave bands (for the observation point at the level of the rail head), it should be noted that in the case of high-speed railway vehicles travelling along a curve, higher values were recorded for bands in the range of 315 Hz to 2000 Hz (difference of 3 to 10.8 dB). In very close proximity to the railway line (5 m), the dominant characteristics are also 80 Hz, 100 Hz, 16,000 Hz and 20,000 Hz, for which the maximum differences with respect to the straight section range from 4.3 dB to 5.3 dB. In case of the observation point at 4 m above rail-head level, individual differences ranging from 3.3 dB to 8.1 dB were recorded. The fairly evenly distributed acoustic pressure levels at 4 m above rail-head level indicate that rolling noise is the dominant noise source for railway vehicles travelling at 200 km/h.

Taking into consideration that more than 97% of standard deviation results—representing the scatter of acoustic pressure levels between the measurement series for a given one-third octave band—did not exceed 2 dB, the average frequency spectra of acoustic pressure levels were determined. In order to compare and present the results of the average spectra of the effect signals, the exposure acoustic pressure levels (dB) were determined for particular ranges of the one-third octave bands. The exposure acoustic pressure levels for the straight section and curve at four measurement points, located at distances of 5 m, 10 m, 20 m and 40 m from the track axis (Figures 10 and 11), were calculated according to the relation [34]:

$$L_{EK} = \left[\frac{1}{n} \sum_{i=1}^{n} 10^{0,1 L_{Eki}} \right] \tag{3}$$

where:

- L_{EK}—exposure level of the acoustic pressure [dB];
- L_{Eki}—level of the acoustic pressure in the n-th one-third octave band [dB].

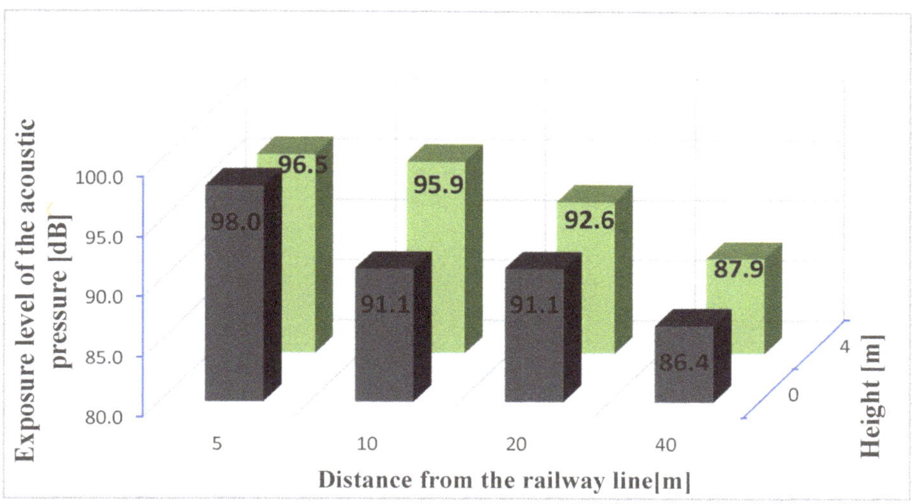

Figure 10. Exposure acoustic pressure level for a straight section at rail-head level (0 m) and at 4 m from rail-head level.

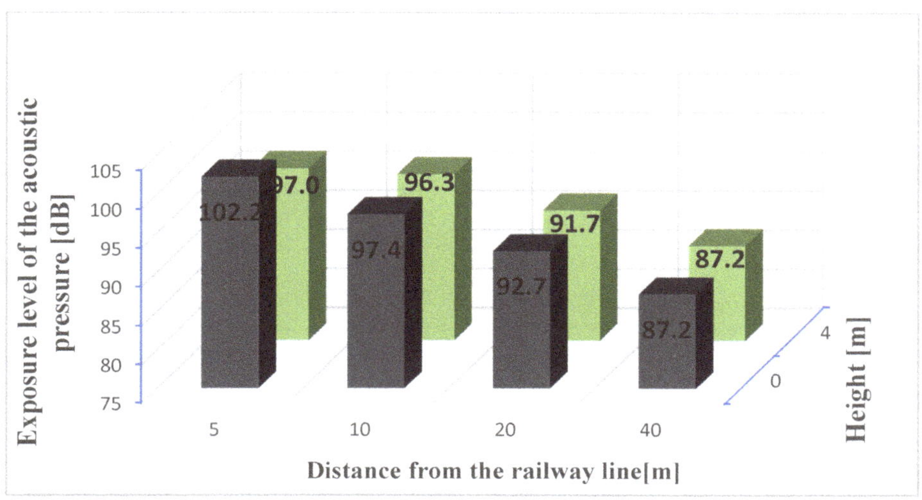

Figure 11. Exposure acoustic pressure level for a curve at rail-head height (0 m) and at 4 m from rail-head height.

The exposure acoustic pressure level at 4 m above rail-head level for high-speed vehicles travelling on a straight section and on a curve is very similar for all distances studied. The differences between the values obtained did not exceed 1 dB, indicating a very similar energy composition in both test cases.

At the level of the rail head (0 m) for the straight section and the curve, at distances of 20 m and 40 m, there were no great differences in the exposure levels (from 0.8 dB to 1.6 dB), which indicates a similar distribution of results. Significant differences in amplitude of exposure acoustic pressure level were recorded between straight and curved section (at the rail-head level) for close distances from the railway line (5 m and 10 m from the track axis), for which the divergence in levels was, respectively, 4.2 dB and 6.3 dB. This situation may be indicative of an increased rolling noise contribution due to the additional friction of the lateral surface of the wheels against the rail.

3.3. Model of Sound Propagation

The construction of the sound-propagation model began with the verification of the batch data. The data obtained from the 4×2 microphone array were verified by determining the correlation coefficient for all 4 s pairs of representative passages. The standard Matlab-*corrcoef* high-level language command was used to determine the correlation coefficients for all the mentioned above pairs. The correlation coefficient is defined as the quotient of the covariance and the product of the standard deviations of the variables under consideration and is determined in accordance with the relation shown in Equation (4) [35,36].

$$P(A, B) = \frac{cov\,(A, B)}{\sigma_A \sigma_B} \tag{4}$$

For about 99% of the cases, the calculated values of the correlation coefficient were not lower than 0.80. The maximum values of the correlation coefficient for the straight section and the curve were equal to 0.998 and 0.999, respectively. In the case of minimum coefficient values, the values were equal to 0.431 for a straight section and 0.753 for a curve. After a preliminary analysis of the obtained data, only the crossings with the highest correlation coefficients were selected for further considerations.

The resulting spectral levels in the one-third octave band of acoustic pressure levels and equivalent continuous sound levels from 10 high-speed rail-vehicle passages were used to determine statistical parameters that provided input data for the propagation models studied. Statistical parameters were defined for each distance from the track centre (i.e., 5 m, 10 m, 20 m, 40 m). The analysis was performed separately for each tested height, i.e., at the level of the rail head and 4 m above the rail head, separately for each measurement cross section (straight section and curve). In order to determine the average level and distribution of values, statistical measures were determined on the basis of representative passages (4 s) in the form of a median, which was calculated on the basis of the relation [35,36]:

$$Me = \begin{cases} \dfrac{x_{N+1}}{2}; & \text{when N is an odd number} \\ \dfrac{x_{\frac{1}{2}N} + x_{\frac{1}{2}N+1}}{2}; & \text{when N is an even number} \end{cases} \tag{5}$$

Based on the obtained batch data, four selected models of railway-noise assessment were analysed in terms of their degree of reflection of the propagation phenomenon. Results of conducted experimental research were used to verify behaviour of particular models describing change of sound level in the *i*-th frequency band as a function of variable distance of observer from the railway line on which high-speed railway vehicles operate.

Four models were verified:

— linear (1st degree);
— using a 2nd degree polynomial;
— using a 3rd degree polynomial;
— power model with a free expression.

An example of noise-propagation model fitting for a straight section described by polynomials of 1st–3rd order (W1R, W2R, W3R), based on experimental data is shown in the following diagrams (Figures 12 and 13).

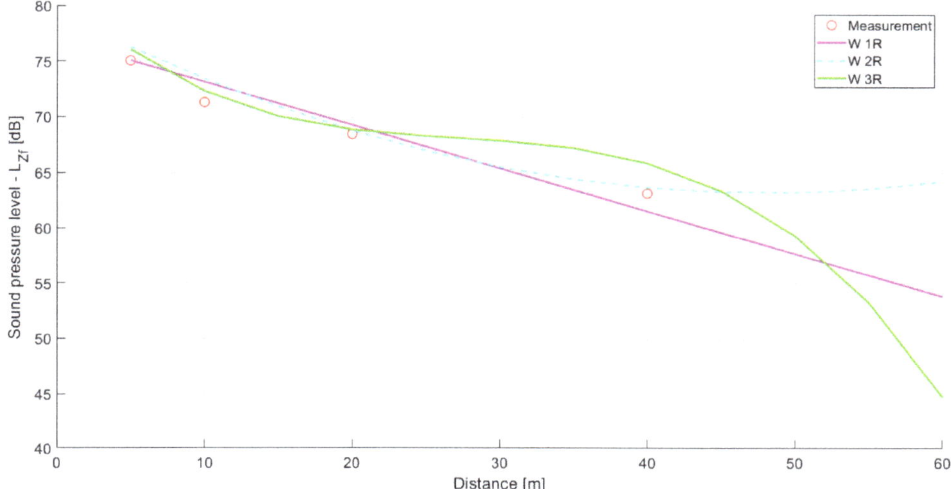

Figure 12. Fitting a 1st–3rd order polynomial noise-propagation model at 250 Hz for a straight section at rail-head level.

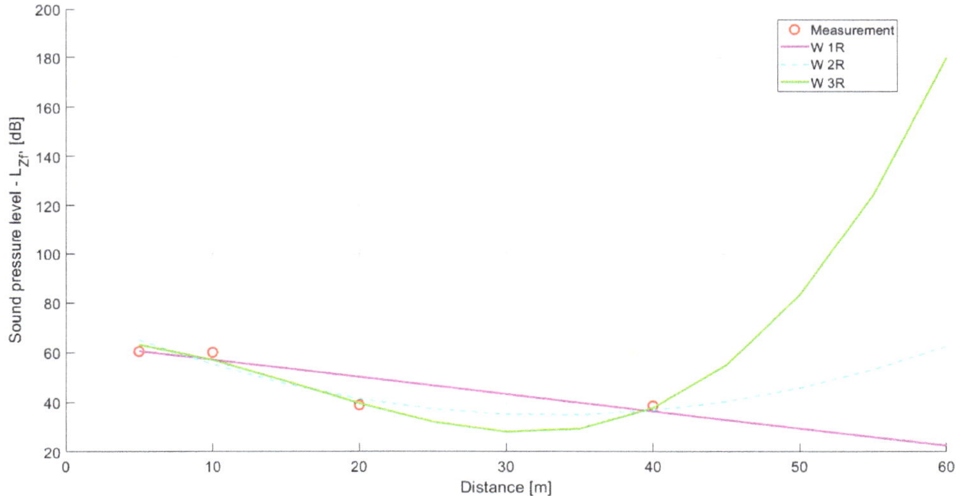

Figure 13. Fitting a 1st–3rd order polynomial noise propagation model at 16,000 Hz for a straight section at rail-head level.

On the basis of the conducted analysis of fitting the noise-propagation model described by polynomials of 1st–3rd order, it may be noticed that all models behave quite well in the range of measurement points (up to 40 m). The obtained characteristics indicate that the linear model, marked in purple on the diagrams, in most cases best represents the results of measurements at specific measurement points, as well as at a distance up to 60 m from the track axis.

The model curves of the 2nd and 3rd degree outside the measurement points are characterised by very large scatter in all ranges of frequency bands and equivalent continuous sound levels. It should be stated that the best representation of the studied phenomena during the passage of railway vehicles at high speeds is a linear model, even though the analysis of characteristics indicates a tendency to too rapid a reduction of noise in the function of distance, beyond the measurement points.

The next step of the study was to verify the representation of the noise-propagation phenomenon using a power model with free expression (FPM). This model takes the form of [37]:

$$L_i = a_i r^{b_i} + c_i \tag{6}$$

where:

- L_i—acoustic pressure level in the i-th frequency band;
- a_i, b_i, c_i—experimental coefficient of the model for t-th frequency band;
- r—distance from sound source.

The final stage of the work was the construction of the author's model using a logarithmic curve.

The evaluation of the tested models showed qualitatively good representation of the measurement results at the selected measurement points. The linear model and the power model with free expression were characterised by the best representation of noise propagation beyond the measurement points. However, they showed a tendency for too rapid a decrease in noise as a function of distance exceeding the distance of the farthest measurement point. Polynomial models of 2nd and 3rd order, outside the measurement points, were characterised by very big errors and unacceptable scattering of results in all ranges of frequency bands and equivalent continuous sound levels.

The author's model, including band noise correction, was also developed and is presented in the study, and was also verified in terms of reproducing the propagation phenomenon. The author's model was characterised by good accuracy for the obtained levels of noise propagation in the range of results obtained experimentally in individual measurement points. Analysis of results obtained for this model—beyond verifiable measurement points, due to the absence of abrupt changes of levels and linear character of decreasing level as a function of further distance—implies high probability of correct representation of the phenomenon of propagation of noise from railway vehicles of high speed by this model. However, it requires further verification in the form of future experimental studies.

Due to the band analysis of the signal in the relation describing the model performance for the i-th frequency band, a correction of the level kpi was introduced for each of the analysed frequency bands (KPM). The relation describing the level for frequency band i at distance x from the sound source is expressed as:

$$L_{xi} = L_{ri} - 10z \log \frac{x}{r} + k_{pi} \tag{7}$$

where:

- L_{xi}—experimentally determined noise level at distance x from the source [dB];
- L_{ri}—sound-level value measured at the reference point r [dB];
- z—source type factor (for a linear source $z = 1$);
- x—distance of observation point from noise source [m];
- r—distance of reference point from noise source [m];
- k_{pi}—level correction for the analysed i-th frequency band [dB].

In order to implement the scheme of the propagation model for each of the analysed frequencies, an attempt was made to determine the parameters of the models by means of the noise-emission correction curve method in Matlab software by using the command-*lsqcurvefit* (with the following parameters: description function, vector of initial conditions, distances of measurement points, values of the sound level in the measurement points). The vector of initial conditions was defined as:

$$x_0 = [s(1)\ s(2)] \tag{8}$$

where:

- $s(1)$—noise level at the reference point, Lri [dB];
- $s(2)$—frequency-band correction value, kpi [dB].

The notation of the function became:

$$fun = @(s,x\_org)\ (s(1) - (10*1*log10(x\_org/5)) + s(2))$$

where x_org was the distance vector of measurement points; therefore, code call *lsqcurvefit* (*fun*,*x0*,*x_org*,*y*) returned model parameters each time. The implementation of the Matlab function code *nlinfit* (with parameters: distance vector of measurement points, level values at measurement points, function, vector of initial conditions) allowed the determination of the residue vector and the Jacobian for the nonlinear regression model returned as a matrix and the estimated variance–covariance matrix for the calculated coefficients, or the calculation of the variance for the calculated error component.

The coefficients of the author's model expressed by Equation (6) for the straight section are shown in Table 2.

Table 2. Coefficient values of the author's noise-propagation model for straight section.

Frequency Band [Hz]	Straight Section			
	Rail-Head Level		4 m over the Rail-Head Level	
	L_{ri}	k_{pi}	L_{ri}	k_{pi}
20	93,787	−2539	93,280	−0881
25	83,886	−1302	80,973	−2334
31.5	84,330	−1372	81,341	−2241
40	85,373	−1833	82,482	−2143
50	86,555	−1622	83,016	−2006
63	87,256	−1503	83,276	−2099
80	86,230	−1306	82,747	−1995
100	85,367	−1520	82,383	−1821
125	83,842	−0822	80,865	−1577
160	81,222	−1256	78,744	−1538
200	78,642	−0850	77,625	−1258
250	76,956	−1113	76,530	−1164
315	77,569	−2319	76,662	−1483
400	77,320	−3356	77,054	−1671
500	76,886	−3900	77,805	−1235
630	77,455	−3988	78,788	−1146
800	78,176	−3703	79,415	−1394
1000	81,087	−3202	81,476	−1173
1250	79,912	−2528	81,934	−0613
1600	81,170	−2677	83,766	−0588
2000	84,022	−2291	86,171	−0287
2500	85,353	−2153	88,165	−0385
3150	86,717	−2800	87,037	−0854
4000	83,389	−2522	84,840	−0648
5000	79,897	−3163	80,953	−1246
6300	77,242	−3355	77,234	−2008
8000	74,420	−3992	74,351	−2188
10,000	73,401	−4356	73,220	−2652
12,500	69,884	−4770	68,740	−2810
16,000	65,911	−5444	63,984	−3596
20,000	61,175	−5888	57,827	−4307
L_{AEq}	58,976	−5511	54,284	−3867

The analysis of the author's model of noise propagation and the free-expression model show that both methods are characterised by an accurate representation of results. The runs of models indicate that the free-expression model more accurately reflects the results of measurements at specific measurement points, but the differences in relation to the author's model are within error limits. Detailed analysis of the obtained characteristics indicates that the author's model is characterised by better runs outside the measurement points. The free-expression power model, outside the range of measurement points, tends in many cases to reduce noise too quickly as a function of distance, which was also confirmed by detailed analyses in the third-order frequency spectra for individual passages (Figures 14–16).

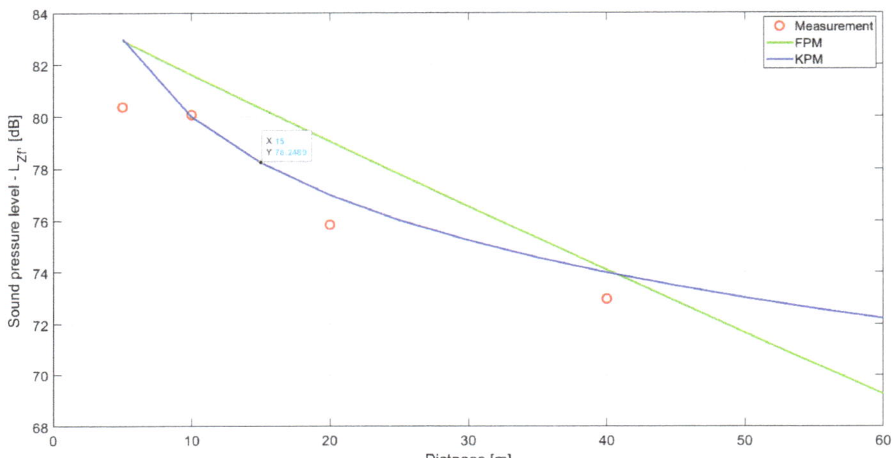

Figure 14. Fitting of the developed noise-propagation model and a free-expression power model for 100 Hz at rail-head level for a straight section.

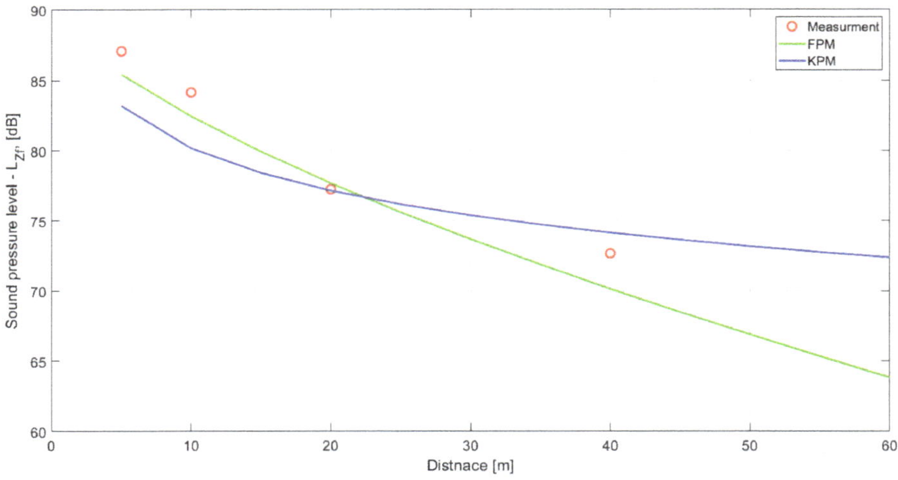

Figure 15. Fitting of the author's noise-propagation model and a free-expression power model for 2000 Hz at rail-head level for a straight section.

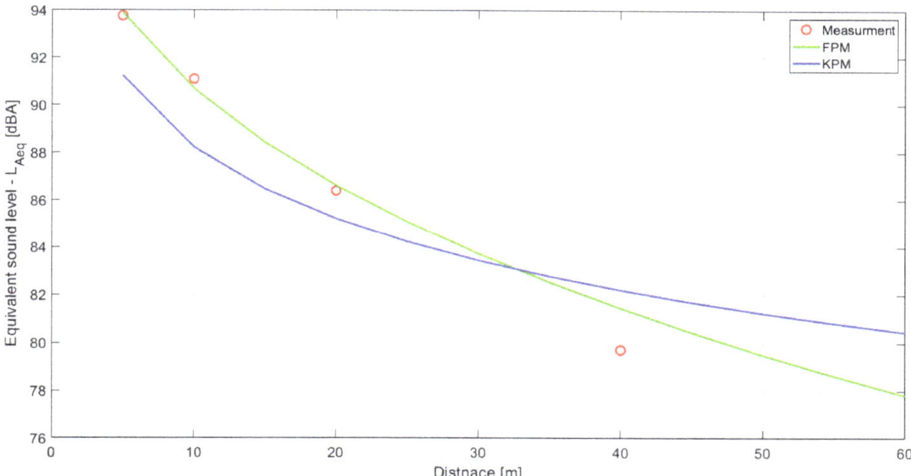

Figure 16. Fitting the author's noise-propagation model and a free-expression power model for the equivalent continuous sound level L_{Aeq}, at rail-head level for a straight section.

4. Conclusions

In the paper, an attempt is made to identify the acoustic signature of a high-speed railway vehicle, moving at a speed of 200 km/h on a straight section and a curve. As part of the experimental research, test fields were determined, measurement equipment was selected, and a methodology for carrying out measurements was specified, including the assessment of noise on a curve and straight line for electric multiple units of Alstom type ETR610-series ED250, the so-called Pendolino. The measurements were made using an acoustic camera and a 4×2 microphone array. As a result of the conducted experimental research, the main sources of noise coming from the studied object were identified and the dominant amplitude–frequency characteristics in the range from 20 Hz to 20 kHz, divided into one-third octave bands, were determined.

Additionally, measurements of the equivalent continuous sound level were made for 20-s passes of high-speed railway vehicles, at all points in the measurement cross section. The conducted experimental research made it possible to compare acoustic phenomena recorded separately for a straight section and a curve. The study confirmed that in the case of electric multiple units travelling at 200 km/h, the dominant noise source is the rolling noise resulting from vibrations at the wheel–rail interface. In the case of high-speed railway vehicles travelling along a curve with a radius of about R = 4000 m, the noise resulting from wheel–rail vibrations does not clearly increase the maximum sound levels compared with a straight section.

The highest acoustic pressure levels from high-speed railway vehicles travelling on a straight section, for heights above rail-head level, were recorded in the low (20 Hz to 160 Hz) and medium (1000 Hz to 4000 Hz) bands. For heights 4 m above rail-head level, the highest levels occurred in the range of 20 Hz to 100 Hz (for a distance of 20 m) and 1000 Hz–4000 Hz.

For vehicles travelling in curves, at rail-head level, the highest values were recorded for frequencies in the low bands from 20 Hz to 100 Hz (for a distance of 40 m) and additionally from 125 Hz to 200 Hz (for 20 m) and in the medium bands from 630 Hz to 3150 Hz (for a distance of 20 m). Additionally, at a distance of 40 m, high levels were recorded at 1600 Hz and 2000 Hz. For a height of 4 m above rail-head level, high values of acoustic pressure levels were recorded in the low bands from 20 to 100 Hz (for a distance of 20 m), and for a distance of 40 m in the range of 50 Hz–100 Hz. High values of sound levels were also recorded for the medium bands in the range from 1000 Hz to 3150 Hz.

On the basis of differences in the measured levels of acoustic pressure for the straight section and the curve, the dominant frequencies in the one-third octave bands (at the level of the rail head) characterising the passage of high-speed railway vehicles along a curve of the radius R = ~4000 m were indicated.

Based on the differences in the measured acoustic pressure levels, the dominant frequencies in the one-third octave bands (at the rail-head level) characterising the passage of high-speed railway vehicles over a curve of radius R = 4000 m were identified. Differences in acoustic pressure levels relative to the straight section were recorded for the bands in the range from 315 Hz to 2000 Hz.

Significant differences in amplitude of exposure acoustic pressure level were recorded between the straight section and the curve (at the level of the rail head) for close distances from railway line (5 m and 10 m from track axis). The divergences from the straight section were 4.2 dB and 6.3 dB, respectively, which may indicate an increased contribution of rolling noise due to the additional friction of the lateral surface of the wheels against the rail head.

In this study, results were developed and presented for the author's model including band noise correction, which was also verified to reproduce the propagation phenomenon. The model was characterised by good accuracy for obtained levels of noise propagation in the range of results obtained experimentally in individual measurement points. The analysis of the results obtained for this model (no abrupt changes of levels, linear character of decreasing levels) indicates that there is a high probability of correct representation of the phenomenon of propagation of noise from high-speed railway vehicles by the author's model, also beyond the measurement points.

Detailed recognition of the acoustic signature of high-speed railway vehicles travelling at 200 km/h will make it possible to minimise the acoustic impact more effectively. By identifying the main sources of noise and knowing the dominant amplitude–frequency characteristics of the studied object, it will be possible to optimally select solutions and measures limiting the negative impact of noise on the environment, including appropriate selection of noise barriers (acoustic screens) aimed at the reduction of particular frequency bands. The methodology presented in this paper to carry out experimental studies based on the measurement of acoustic phenomena using an acoustic camera and microphone matrix can also be used for other railway vehicles. Due to the differences in speed and shape of the locomotive/traction unit, it should be assumed that the recognised acoustic signal will have different amplitude and frequency characteristics. Further work directions include carrying out acoustic measurements for other railway vehicles in order to verify and compare characteristics of particular research objects.

Author Contributions: Conceptualisation, K.P. and J.K.; methodology, J.K.; validation, J.K. and K.P.; formal analysis, J.K. and K.P.; investigation, K.P.; resources, K.P.; data curation, K.P.; writing—original draft preparation, K.P.; writing—review and editing, J.K.; visualisation, K.P.; supervision, J.K.; project administration, K.P. All authors have read and agreed to the published version of the manuscript.

Funding: This research received no external funding.

Institutional Review Board Statement: Not applicable.

Informed Consent Statement: Not applicable.

Conflicts of Interest: The authors declare no conflict of interest.

References

1. Peris, E. *Environmental Noise in Europe—2020, European Environment Agency Raport*; European Union: Luxembourg, 2020; No 22/2019.
2. Polak, K.; Korzeb, J. Identification of the major noise energy sources in rail vehicles moving at a speed of 200 km/h. *Energies* **2021**, *14*, 3957. [CrossRef]
3. Polak, K.; Korzeb, J. Measurements of noise from high-speed rail vehicles, Railway Problems. Pomiary hałasu pochodzącego od pojazdów kolejowych zwiększonych prędkości. *Probl. Kolejnictwa* **2019**, *184*, 33–38.
4. Żurkowski, A. Central Trunk Line (CMK) 200km/h. Traffic, timetable and operations. *Tech. Transp. Szyn.* **2015**, *nr 6/2015*, 24–25.

5. Graf, H.R.; Czolbe, C. Pass-by noise source identification for railroad cars using array measurements. In Proceedings of the 23rd International Congress on Acoustics, Aachen, Germany, 9–13 September 2019; pp. 1574–1581.

6. Jiang, S.; Yang, S.; Wu, D.; Wen, B.-C. Prediction and validation for the aerodynamic noise of high-speed train power car. *Transp. Probl.* **2018**, *13*, 91–102. [CrossRef]

7. Li, L.; Thompson, D.; Xie, Y.; Zhu, Q.; Luo, Z.; Lei, Z. Influence of rail fastener stiffness on railway vehicle interior noise. *Appl. Acoust.* **2019**, *145*, 69–81. [CrossRef]

8. Maillard, J.; Van-Maercke, D.; Poisson, F.; Regairaz, J.-P.; Dufour, J.B. Comparison of pass-by railway noise indicators obtained from standard engineering methods with measured values. In Proceedings of the Forum Acusticum, Lyon, France, 7–11 December 2020; pp. 2477–2483. [CrossRef]

9. Uda, T.; Kitagawa, T.; Saito, S.; Wakabayashi, Y. Low frequency aerodynamic noise from high speed trains. *Q. Rep. RTRI* **2018**, *59*, 109–114. [CrossRef]

10. Zea, E.; Manzari, L.; Squicciarini, G.; Feng, L.; Thompson, D.; Arteaga, L.I. Wavenumber–domain separation of rail contributionto pass-by noise. *J. Sound Vib.* **2017**, *409*, 24–42. [CrossRef]

11. Zhang, J.; Xiao, X.; Wang, D.; Yang, Y.; Fan, J. Source contribution analysis for exterior noise of a high-speed train: Experiments and simulations. *Shock. Vib.* **2018**, *2018*, 5319460. [CrossRef]

12. Zhao, C.; Wang, P.; Yi, Q.; Sheng, X.; Lu, J. A detailed experimental study of the validity and applicability of slotted stand-off layer rail dampers in reducing railway vibration and noise. *J. Low Freq. Noise Vib. Act. Control.* **2018**, *37*, 896–910. [CrossRef]

13. *EN ISO 3095*; 2013 Acoustics—Railway Applications—Measurement of Noise Emitted by Railbound Vehicles. CEN: Brussels, Belgium, 2013.

14. Li, M.; Zhong, S.; Deng, T.; Zhu, Z.; Sheng, X. Analysis of source contribution to pass-by noise for a moving highspeed train based on microphone array measurement Measurement. *J. Int. Meas. Confed.* **2021**, *174*, 109058. [CrossRef]

15. Sheng, X.; Cheng, G.; Thompson, D.J. Modelling wheel/rail rolling noise for a highspeed train running along an infinitel long periodic slab track. *J. Acoust. Soc. Am.* **2020**, *148*, 174–190. [CrossRef] [PubMed]

16. Zhang, J.; Xiao, X.; Sheng, X.; Li, Z. Sound source localisation for a high-speed train and its transfer path to interior noise. *Chin. J. Mech. Eng.* **2019**, *32*, 1–16. [CrossRef]

17. Wawrzyniak, A. ETR610 electric multiple units of the ED250 series for PKP Intercity, S.A. Elektrycznepoci ągizespołowe ETR610 serii ED250 dla PKP Intercity S.A. *Tech. Transp. Szyn.* **2013**, *9*, 20–24. (In Polish)

18. Pachla, F. Transmission of vibrations from the ground to the foundation from the movement of passenger trains. *Vibroengineering Procedia* **2019**, *27*, 93–96. [CrossRef]

19. Graff, M. The New Electric Multiple Units for Long-Distance and Regional Traffic in Poland in 2015. *Tech. Transp. Szyn.* **2016**, *nr 1-2/2016*, 22–33.

20. Czarnecki, M.; Witold, G.; Massel, A.; Walczak, S. Introduction of High-Speed Rolling Stock into Operation on the Polish Railway Network. In *High-Speed Rail in Poland: Advances and Perspectives*; Zurkowski, A., Ed.; CRC Press: Warsaw, Poland, 2018; pp. 203–228. [CrossRef]

21. Raczyński, J. The ED250 (Pendolino) Experiences in Poland—First Years of Exploitation, Technika Transportu Szynowego. 2018. no 4/2018. pp. 30–32. Available online: http://yadda.icm.edu.pl/yadda/element/bwmeta1.element.baztech-9b7e36f1-647f-4ee4-aaf5-b0e7d9495dcd/c/Raczynski1TTS4en.pdf (accessed on 22 February 2022).

22. Id-12 (D-29). The List of Lines, Introduced by Order No. 1/2009 of the Management Board of PKP PolskieLinieKolejowe, S.A. of 9 February 2009, as AMENDED., PKP Polskie Linie Kolejowe. 2009. Available online: https://www.plk-sa.pl/files/public/user_upload/pdf/Akty_prawne_i_przepisy/Instrukcje/Wydruk/Id-12_Wykaz_linii_06.2021.pdf (accessed on 22 February 2022). (In Polish).

23. Gade, S.; Ginn, B.; Gomes, J.; Hald, J. Recent advances in rail vehicle moving source beamforming. Instrumentation Reference. *Sound Vib.* **2015**, *4*, 8–14.

24. Le Courtois, F.; Thomas, J.-H.; Poisson, F.; Pascal, J.-C. Genetic optimisation of a plane array geometry for beamforming. Application to source localisation in a high speed train. *J. Sound Vib.* **2016**, *371*, 78–93. [CrossRef]

25. Meng, F.; Wefers, F.; Vorlaender, M. Acquisition of exterior multiple soundsources for train auralization based on beamforming. In Proceedings of the the the 10th European Congress and Exposition on Noise Control Engineering, Maastricht, The Netherlands, 31 May–3 June 2015; pp. 1703–1708.

26. Go, Y.-J.; Choi, J.-S. An acoustic source localization method using a drone-mounted phased microphone array. *Drones* **2021**, *5*, 75. [CrossRef]

27. Zhang, J.; Squicciarini, G.; Thompson, D.J. Implications of the directivity of railway noise sources for their quantification using conventional beamforming. *J. Sound Vib.* **2019**, *459*, 114841. [CrossRef]

28. Chiariotti, P.; Martarell, M.; Castellini, P. Acoustic beamforming for noise source localization—reviews, methodology and applications. *Mech. Syst. Signal Process* **2019**, *120*, 422–448. [CrossRef]

29. Polak, F.; Sikorski, W.; Siodła, K. Prototype measurement system for locating sources of acoustic emission-microphone matrix. Poznan University of Technology Academic Journals. *Electr. Eng.* **2015**, *84*, 207–214. (In Polish)

30. Aczel, A.D. *Statistics in Management*, 2nd ed.; PWN: Warszawa, Poland, 2018. (In Polish)

31. Ozimek, E. Sound and its perception. In *Physical and Psychoacoustic Aspects*; PWN: Warszawa, Poland, 2018. (In Polish)

32. Sobczyk, M. *Statistic*; PWN: Warszawa, Poland, 2021; EAN: 9788301151997. (In Polish)

33. Kokowski, P.; Laboratory of Applied Acoustics. *Theoretical Introduction*; UAM: Poznań, Poland, 2003. (In Polish)
34. Ordinance of the Minister of the Environment of 16 June 2011 on the requirements for carrying out measurements of the levels of substances or energy in the environment by the manager of a road, railway line, tramway line, airport or port. *J. Laws* **2011**, *140*, 824.
35. Bobowski, Z. *Selected Methods of Descriptive Statistics and Statistical Inference*; WWSZiP: Wałbrzych, Poland, 2004. (In Polish)
36. Muciek, A. *Determination of Mathematical Models from Experimental Data*; Oficyna Wydawnicza Politechniki Wrocławskiej: Wrocław, Poland, 2012. (In Polish)
37. Borwin, M.; Komorski, P.; Szymański, G. Modeling of the sound propagation for a railway vehicle moving on the track. *Pojazdy Szyn.* **2019**, *3*, 60–68. [CrossRef]

Article

Synthesis of the Current Controller of the Vector Control System for Asynchronous Traction Drive of Electric Locomotives

Sergey Goolak [1], Viktor Tkachenko [1], Svitlana Sapronova [2], Vaidas Lukoševičius [3,*], Robertas Keršys [3], Rolandas Makaras [3], Artūras Keršys [3] and Borys Liubarskyi [4]

[1] Department of Electromechanics and Rolling Stock of Railways, State University of Infrastructure and Technologies, Kyrylivska Str., 9, 04071 Kyiv, Ukraine; gulak_so@gsuite.duit.edu.ua (S.G.); v.p.tkachenko@gsuite.duit.edu.ua (V.T.)

[2] Department of Cars and Carriage Facilities, State University of Infrastructure and Technologies, Kyrylivska Str., 9, 04071 Kyiv, Ukraine; sapronova_sy@gsuite.duit.edu.ua

[3] Faculty of Mechanical Engineering and Design, Kaunas University of Technology, Studentų Str., 56, 51424 Kaunas, Lithuania; robertas.kersys@ktu.lt (R.K.); rolandas.makaras@ktu.lt (R.M.); arturas.kersys@ktu.lt (A.K.)

[4] Department of Electrical Transport and Diesel Locomotive, National Technical University "Kharkiv Polytechnic Institute", Kyrpychova Str., 2, 61002 Kharkiv, Ukraine; borys.liubarskyi@khpi.edu.ua

* Correspondence: vaidas.lukosevicius@ktu.lt

Abstract: This paper deals with the analysis of the operating conditions of traction drives of the electric locomotives with asynchronous traction motors. The process of change of the catenary system voltage was found to have a stochastic character. The method of current controller synthesis based on the Wiener–Hopf equation was proposed to enable efficient performance of the traction drive control system under the condition of the stochastic nature of the catenary system voltage and the presence of interferences, when measuring the stator current values of the tractor motor. Performance simulation of the proposed current controller and the current controller used in the existing vector control systems of the traction drives used in the electric locomotives was implemented. The results of the performance simulation of the proposed current controller were compared with the performance of the current controller in existing vector control systems of the traction drives. The results are applicable to the design of vector control systems of traction drives in electric locomotives and to the study of the influence of performance of electric traction drives in electric locomotives on the quality indicators of the power supplied by the traction power supply system under the actual operating conditions of the locomotive.

Keywords: optimal controller; traction drive; vector control system

Citation: Goolak, S.; Tkachenko, V.; Sapronova, S.; Lukoševičius, V.; Keršys, R.; Makaras, R.; Keršys, A.; Liubarskyi, B. Synthesis of the Current Controller of the Vector Control System for Asynchronous Traction Drive of Electric Locomotives. *Energies* **2022**, *15*, 2374. https://doi.org/10.3390/en15072374

Academic Editors: Larysa Neduzha, Jan Kalivoda and Abu-Siada Ahmed

Received: 23 February 2022
Accepted: 22 March 2022
Published: 24 March 2022

Publisher's Note: MDPI stays neutral with regard to jurisdictional claims in published maps and institutional affiliations.

Copyright: © 2022 by the authors. Licensee MDPI, Basel, Switzerland. This article is an open access article distributed under the terms and conditions of the Creative Commons Attribution (CC BY) license (https://creativecommons.org/licenses/by/4.0/).

1. Introduction

The selection of an optimal control option for specific technical applications should be based on the reasonably selected methods appropriate for the respective objects. In terms of traction drive control in an AC electric locomotive, the choice should account for a series of potential uncertainties. These uncertainties are primarily related to the operating conditions of AC electric locomotives. Studies [1–3] have shown that the voltage change process in the catenary system of the AC traction power supply network is a stochastic process. The stochastic character of the voltage change process in the catenary system of the traction power supply network is determined by the conditions under which the electric rolling stock passes through the feeder zone [4,5], the electromagnetic compatibility of the units of the electric rolling stock that are located in a single feeder zone at the quality of same time [3,6], and the current collection quality [7–9].

It should also be noted that the load of the traction drive in an electric rolling stock changes as the wheel of the electric rolling stock interacts with the rail. Several studies have been dedicated to this problem. The study [10] investigates the influence of curved line track spans on the traction drive load. The influence of the track profile on the characteristics of the traction drive has been demonstrated in studies [11,12]. The investigation of the wheel-rail friction coefficient can be found in the study [13]. An analysis of studies dedicated to the interaction of the influence of the electric rolling stock (ERS) wheel on the characteristics of the traction drive has suggested that the change in load of the electric rolling stock traction drive also has a stochastic character. In summary, an analysis conducted on the influence of the operating conditions of the ERS on the electromechanical processes in the traction drive has suggested the following conclusion: both the change of voltage in the catenary system and the change of the traction drive load have a stochastic character. The stochastic character of the variation in the voltage of the catenary system and the moment of load of the traction drive leads to greater losses in the traction motors [14] and affects the stability of performance of the traction drive control system [15–17].

In view of the stochastic character of variation of the voltage parameters of the catenary network and the traction drive load, it would be particularly practical to analyze the electromechanical processes in the traction drive system under the actual operating conditions of the AC electric locomotives. Practicality is primarily related to the fact that the above factors act on the quality of drive control. The quality of drive control, in turn, is associated with losses in the drive, which act on the energy efficiency of the performance of the traction drive system. The control system is responsible for the quality of control in the traction drive system. An analysis of traction drive systems in AC electric locomotives has suggested that asynchronous motors are the most widely used traction motors in contemporary AC electric locomotives [18]. Study [18] also provides the topologies for the design of traction drive control systems using the asynchronous motor. The control systems for this kind of drive may be the following: scalar, vector, and direct torque control (DTC).

The scalar control system belongs to the category of design-free open-loop control systems [19,20]. Open-loop control systems do not offer any possibilities for optimization of the control algorithm. Optimal control systems are designed on the basis of closed-loop control systems [21–23].

DTC systems [24–26] and vector control systems [27–29] belong to the category of design-closed-loop automatic control systems. This category of systems offers the possibility to apply the principles of optimal control [21–23].

The vector control systems of traction drives have become the most widespread systems in rolling stocks [18]. As a result, a major focus is placed on vector control systems. Comparison of the strengths and drawbacks of vector control systems versus DTC [27–29] has suggested that, in contrast to DTC systems, the drawbacks of vector control systems may include slow response to change in the resistance torque of the traction drive. In light of this drawback of vector control systems, the stochastic character of the change in the traction drive load can be further ignored. Hence, optimization of performance of the vector control system should account for the stochastic character of only the process of change of voltage in the catenary system.

Two directions can be identified in the design of the optimal control systems: optimization of performance of the overall system, and optimization of individual elements of the system, i.e., the controllers. Traction drive control systems are designed on the basis of the energy efficiency criterion [30,31].

In terms of energy efficiency, the Pontryagin criterion is the most effective when designing optimal control systems [32,33]. However, the design of the control system that is optimal by Pontryagin's criterion has certain implications. The implications stem from the double-channel character of the vector system. Here, the control is implemented using two channels: flux linkage channel and speed channel. It should also be noted that the control using the two channels is distributed by time: the flux linkage control channel is the first to operate, and the speed channel is activated after the former has completed its operation [27–29]. This entails certain difficulties in the practical implementation of this kind of system [34].

This issue can be solved by analytic design of optimal controllers. Proportional integral controllers (PI) are used as current controllers in electric locomotives with vector control systems. The parameters of this kind of controller are based on the design of the desired logarithmic amplitude-frequency characteristic of the current control circuit [27–29]. This approach can be considered appropriate in view of the hypothesis that the process of change of the stator currents in an asynchronous motor is deterministic. As demonstrated above, this process is stochastic under the actual operating conditions of electric locomotives. Furthermore, another factor that influences the quality of control has not been taken into account in the design of this kind of controllers in these systems [27–29]. It is related to the following circumstances. In a vector control system, current sensors are activated in the feedback of the current control channel. Sensors enable identification of the present values of the phase currents of the asynchronous motor stator. Electromagnetic interference of the current sensors is caused by the asynchronous traction motor. These interferences are usually so-called 'white noise', that is, they have zero mathematical expectation and nonzero dispersion [35].

The study [36] is dedicated to solving the problem of synthesis of the optimal current controller operating under stochastic conditions [36]. Despite the advantage of this approach, namely, the simplicity of practical implementation of the algorithm, the question of control quality remains open. The controller parameters are calculated on the basis of the energy efficiency criterion for the specified design of the controller. However, they are optimal for the specified design only. Hence, this is not necessarily an optimal solution in terms of the quality of the control.

This weakness can be eliminated by synthesis of the optimal controller having an arbitrary design [37]. In case of this approach to the design of optimal controllers, it is the controller design that is subject to optimization, and its parameters are calculated specifically for the resulting design of the controller. However, this approach to the design of optimal controllers leads to uncertainty. It is related to the necessity of solving a system of equations where there are more unknowns than the equations.

The study [38] is dedicated to the synthesis of the optimal controller using the maximum principle. For the controllers based on the criterion of maximum, the design is synthesized, and the optimal controller parameters are calculated. Despite the obviously correct approach to solving the task of synthesis of the optimal controller, the algorithm specified has an essential drawback. The drawback is related to the fact that at low values of the error signal at the input, the output signal of the controller changes according to the linear law. The signal reaches the maximum possible value at the controller output, i.e., the controller enters the saturation mode under the presence of significant values of the error signal at the input.

This drawback may be eliminated by synthesis of the optimal controller using the dynamic programming method [39]. This algorithm is effective for the design of optimal controllers, provided that there are no control-related limitations. In case of any limitations, the method becomes difficult to implement. This is due to the fact that non-zero initial conditions need to be entered into the system of differential equations describing the system performance. This prompts the necessity to solve the system of non-homogenous differential equations.

Optimal filtration methods may be used for the synthesis of optimal controllers under stochastic conditions. In the case of the deterministic control signal and stochastic interference, the Kalman–Bucy filter method is the most effective [40]. This filter is implemented for the purpose of writing down the transfer function of the system by using the state-space representation. It is the most effective when analyzing the performance of the system in the time domain.

However, vector systems of drive control in asynchronous motors are discrete. They are implemented by means of z-transformation, i.e., in the frequency domain. In the frequency domain, the controllers synthesized by using the Wiener–Hopf equation are the most effective. These controllers are also the most effective under the conditions of stochasticity of both the control signal and the interference [41,42].

Hence, it is relevant to implement the synthesis of optimal current controller on the basis of the Wiener–Hopf equation for the vector control system for the traction drive in electric locomotives with asynchronous motors.

The purpose of the study was to perform the synthesis of an optimal current controller based on the Wiener–Hopf equation for the vector control system for traction drive in electric locomotives with asynchronous motors.

The following was implemented for the purpose of the specified objective:

- The structural scheme was obtained in the transfer functions of the control channel for the current of the vector control systems for the traction drives in the AC electric locomotives;
- The synthesis of design and parameters of optimal current controller was obtained by using the Wiener–Hopf equation for the vector control system for the traction drive in the electric locomotive with asynchronous motors;
- Modeling of the synthesized current controller performance under the condition of deterministic character of the control signal and stochastic 'white-noise'-type interference;
- Comparison of the resulting transient characteristic of the synthesized controller with the transfer characteristic of the PI controller that is used in vector control systems for the traction drive in the AC electric locomotives.

The study may be used when investigating the energy efficiency of traction drives in AC electric locomotives, the interaction between the catenary system and the ERS, and the optimization of the control systems for traction drives in electric locomotives.

2. Materials and Methods

The optimal current synthesis was performed for the vector control system for the traction drive with the asynchronous motor. The structural scheme of the vector control system is presented in the study [28]. The control scheme parameters are presented in relative units.

A structural scheme in the transfer functions was designed for the vector control system. Current control channels were identified in the scheme. Optimal current controllers were synthesized for the current control channels by using the Wiener–Hopf equation.

A simulation model of the current control channel with the synthesized current controller was developed in the MATLab software environment. The transient characteristic of the current control channel was obtained for the conditions where the deterministic signal acted as a control signal at the controller input, and the stochastic 'white-noise'-type signal acted as an interference.

The transient characteristic was obtained for the initial control scheme for the same initial conditions. In this scheme, the PI controller was used as the current controller.

The transient characteristics obtained were compared in terms of the errors of the output control signals. The model was performed for the steady-state operation of the traction drive under its nominal operating conditions.

An AC electric locomotive with series DC-3 (ДС-3) asynchronous traction motor (Dnipro, Ukraine) was chosen as the object of the investigation. Type AD914U1 (АД914У1) asynchronous motors are used as traction motors in the electric locomotives of this series. Technical specifications of the AD914U1 (АД914У1) traction motor are presented in Table 1 [43].

Table 1. Parameters of AD914U1 (АД914У1) traction motor.

Parameter	Value
Power P, kW	1200
Phase-to-phase RMS voltage U_{nom}, V	1870
RMS value I_{nom}, A	450
Rated frequency of the supply voltage f, Hz	55.8
Number of phases n, pcs.	3
Number of pole pairs p_p	3
Nominal rotational speed n_r, rpm	1110
Efficiency η, %	95.5
Power factor $\cos\varphi$, per unit	0.88
Active resistance of the stator winding r_s, Ω	0.0226
Active resistance of the rotor winding reduced to the stator winding r'_r, Ohm	0.0261
Stator winding leakage inductance $L_{\sigma s}$, Hn	0.00065
Rotor winding leakage inductance reduced to the stator winding, $L'_{\sigma r}$, Hn	0.00045
Total inductance of the magnetizing circuit L_{μ}, Hn	0.0194336
Moment of inertia of the motor J, kg·m^2	73

Current controller calculations were performed using the methodology presented in the study [44]. Calculations were performed using relative units. The calculation results are presented in Table 2.

Table 2. Calculation results for the basic values and parameters of the controller.

Parameter	Designation	Value
Current loop tuning coefficient X, r.u.	a_{Ix}	2
Current loop tuning coefficient Y, r.u.	a_{Iy}	2
Proportional coefficient of current controller X, r.u.	K_{pIx}	0.155
Integral coefficient of current controller X, r.u.	K_{iIx}	0.00922
Proportional coefficient of current controller Y, r.u.	K_{pIy}	0.155
Integral coefficient of current controller Y, r.u.	K_{iIy}	0.00922

The remaining parameters necessary for the synthesis of an optimal current controller were obtained as a result of the calculations and are provided below.

3. Calculations of the Optimal Current Controller Parameters by Using the Wiener–Hopf Equation

3.1. Definition of the Current Control Channels in the Transfer Functions of the Vector Control System

The structural scheme of the vector control system is presented in the study [28] and is depicted in Figure 1.

The following assumptions were made during the development of the above structural diagram:

1. In this paper, research on system operation of asynchronous motor asymmetric modes and on asynchronous motor asymmetric power supply system will not be conducted. Therefore, the induction motor model is chosen in x–y coordinates in order to avoid coordinate conversion;

2. The investigation will be carried out in steady-state mode of operation at motor shaft rotation frequency $\omega = \omega_{set}$. For this mode, the stator currents are fixed: $I_x^* = 1$ and $I_y^* = 0$, respectively.

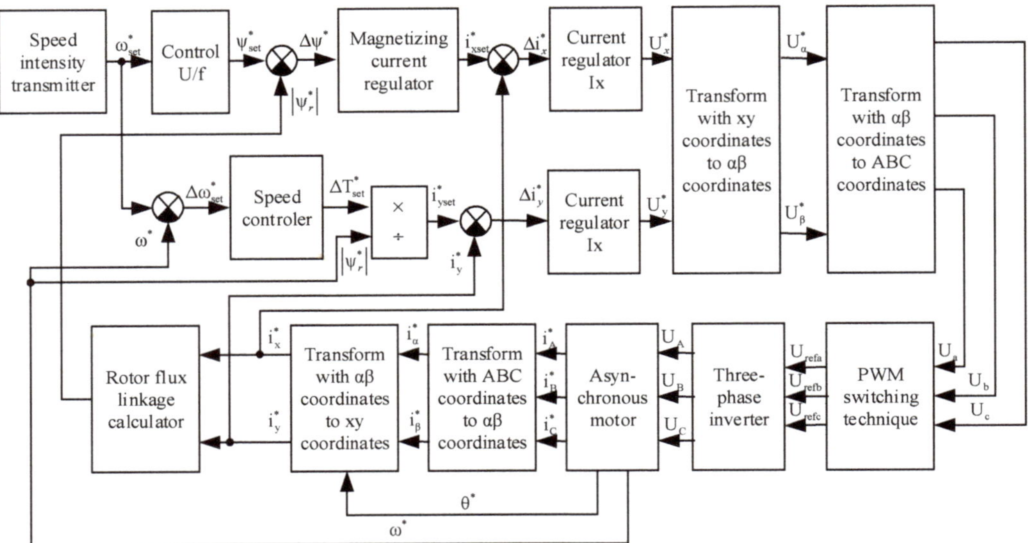

Figure 1. Structural scheme of the vector control system (* values presented in relative units of measurement).

Figure 1 depicts the structural scheme of the vector control system in the case of the common system of coordinates α-β oriented according to the flux linkage of the rotor in the system. Therefore, to define the structural scheme of the current control channel, the structural scheme of the 'Frequency converter—Asynchronous motor' system should be developed for the common system of coordinates α-β, which are oriented according to the flux linkage of the rotor.

Taking into account the recommendations for creating a vector control system with an induction motor represented in x–y coordinates [45], the structural diagram of the vector control system for an induction motor (Figure 1) is presented in transfer functions (Figure 2).

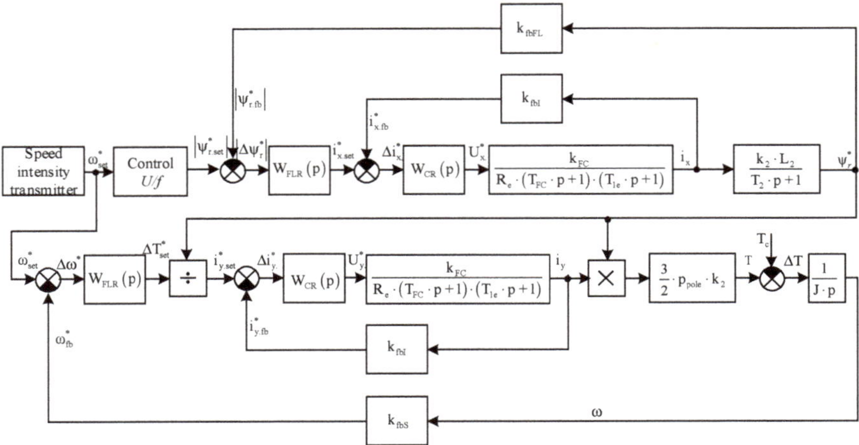

Figure 2. The structural scheme of the system 'Frequency converter—asynchronous motor' in the case of the common coordinate system of coordinates α-β, which are oriented according to the flux linkage of the rotor.

Figure 2 includes the following designations:

k_{fbI}, k_{fbFL}, and k_{fbS}—feedback coefficients for the current, flux linkage, and speed, respectively;
T_{1e}, T_2, and T_{FC}—time constants for the stator circuit, rotor circuit, and the frequency converter, respectively;
$W_{CR(p)}$, $W_{FL(p)}$, and $W_{SR(p)}$—transfer functions of the current controller, flux linkage controller, and speed, respectively;
k_2—coefficient of the electromagnetic link of the rotor;
k_{FC}—coefficient of the transfer function of the frequency converter;
R_{1e}—equivalent active resistance of the stator circuit.

The values indicated with the index * are used in relative units.

To synthesize a current regulator, a current transfer channel with a transfer $W(p)_C = i(p)/i_{set}(p)$ function for x and y coordinates should be identified. Figure 2 suggests that the transferred functions of the current control channels are the same for coordinates x–y. Given the assumption that studies are conducted for the mode $\omega = \omega_{set}$, for which the stator currents are fixed: $I_x^* = 1$ and $I_y^* = 0$ the structural diagram of the current transmission channel is shown in Figure 3.

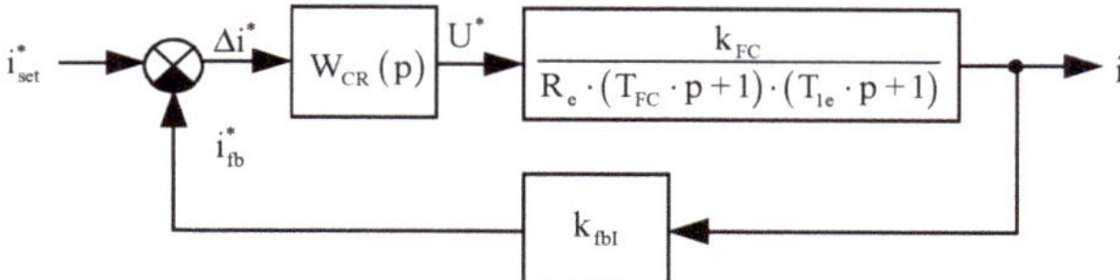

Figure 3. Structural scheme of the current control channels (* values presented in relative units of measurement).

For convenience of subsequent calculations, the following transformations of the structural scheme were performed (Figure 3):

— The scheme (Figure 3) with a non-unique transfer function of feedback (k_{fbI}) was transformed into the structural scheme with a unique feedback connection (Figure 4);
— In the scheme (Figure 4), the interferences acting on the current sensors in the stator circuit was depicted as signal f;
— The signal having the specified stator current value (I_{1set}) was designated as x, and the output signal of stator current (I_1) was designated as y.

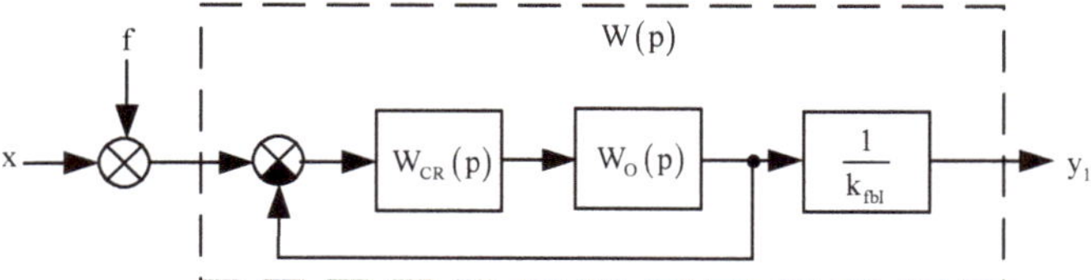

Figure 4. Design structural scheme of the current control channels.

The design scheme for the current control channel resulting from the transformations above is presented in Figure 4.

Transfer function of the control object in the design scheme:

$$W'_O(p) = \frac{k_{fbI} \cdot k_{fFC}}{R_{e1} \cdot (T_{FC} \cdot p + 1) \cdot (T_{e1} \cdot p + 1)} \qquad (1)$$

In expression (1), active resistance of the stator circuit was defined by expression:

$$R_{e1} = r_s + r'_r \cdot k_2^2 \qquad (2)$$

where $r_s = 0.0226\ \Omega$—active resistance of the stator phase (Table 1); $r'_r = 0.0261\ \Omega$—active resistance of the rotor phase reduced to the stator (Table 1); $k_2 = L_\mu/(L'_{\sigma r} + L_\mu) = 0.9774$—coefficient of the electromagnetic link of the rotor; $L_\mu = 0.0194336$ Hn—inductance of the magnetic circuit of the motor (Table 1); and $L'_{\sigma r} = 0.00045$ Hn—inductance of the rotor phase leakage reduced to the stator (Table 1).

The time constant of the stator circuit in Equation (1) was determined by using formula:

$$T_{e1} = \frac{k_2 \cdot L_{\sigma e}}{R_{e1}} = 0.023 \text{ s} \tag{3}$$

where $L_{\sigma e}$—equivalent inductance of the motor leakage.

$$L_{\sigma e} = L_{\sigma s} + L'_{\sigma r} + \frac{L_{\sigma s} \cdot L'_{\sigma r}}{L_\mu} = 0.001115 \text{ Hn} \tag{4}$$

where $L_{\sigma s}$—inductance of the stator winding leakage.

Time constant of the frequency converter in Equation (1) was determined by using formula:

$$T_{FC} = a_T \cdot T_{e1} = 2 \times 0.023 = 0.046 \text{ s} \tag{5}$$

where a_T—current controller tuning coefficient. It is usually accepted that $a_T = 2$.

The transfer coefficient of the frequency converter is determined by expression:

$$k_{FC} = \frac{I_{snom}}{I_1} = \frac{450}{1} = 450 \text{ A} \tag{6}$$

where $I_{snom} = 450$ A—nominal value of the stator current (Table 1) and I_1—nominal current value at the current controller output. The vector control scheme (Figure 1) was implemented in relative units, $I_1 = 1$ A.

The feedback coefficient for the current:

$$k_{fbI} = \frac{1}{I_{snom}} = \frac{1}{450} = 0.0022 \text{ 1/A} \tag{7}$$

The transfer scheme of the current control circuit is therefore the following:

$$W(p) = \frac{W'_O(p) \cdot W_{FC}(p)}{1 + W'_O(p) \cdot W_{FC}(p)} \cdot \frac{1}{k_{fbI}} = \frac{k_{fFC}}{1 + k_{fbI} \cdot R_{e1} \cdot W'_O(p) \cdot W_{FC}(p)} \tag{8}$$

To determine the structure and parameters of the optimal current controller using the Wiener–Hopf equation, it is necessary to determine the signal parameters and the interferences that act at the controller input (Figure 4).

3.2. Determination of the Parameters of the Signal and the Interferences That Act at the Controller Input

Reference signal $x(t) = \alpha \cdot \sin(\omega \cdot t + \beta)$ and 'white-noise'-type interference $f(t)$ act at the system input (Figure 4) [46]. In the performed analysis, it was found that vector drive control systems with asynchronous motors are discrete systems. They are implemented using the z-transformation, i.e., in the frequency domain. In the frequency domain, the most effective regulators are those synthesized using the Wiener–Hopf equation. Additionally, the regulators synthesized with the Wiener–Hopf equation are the most effective under the condition of stochasticity of both the control signal and the disturbance [41,42].

According to the methodology for the calculation of the optimal controller using the Wiener–Hopf equation [41,42], it is necessary to obtain the spectral densities of the control signal and of the interference. To determine the spectral density of the signal at the controller input $x(t)$ taking into account the structure of the vector control system, the following operations had to be performed:

- Time dependences of the phase currents of the stator had to be obtained for the model;
- The resulting dependences had to be transformed for the system of coordinates x–y;
- The signal parameters had to be calculated.

Calculation of the controller parameters is performed for the steady-state nominal operation of the control system. As a result, to reduce the number of coordinate transformations, the following hypothesis was established: it can be assumed, with a certain minor error, that the current at the input of the current control channel (Figure 3) is equal to phase A stator current of the traction motor. Studies [43,47] propose the model of АД914У1 asynchronous motor in three-phase coordinates. This model provided the time diagrams of the stator current and phase voltage of phase A for the steady-state mode (Figure 5a). These time diagrams were used to determine the parameters of signal x(t): amplitude $\alpha = 450$ A, phase shift between the voltage and the current $\beta = 1.4$ rad. Using this model, time diagrams of the stator current and phase voltage of phase A for steady state mode were obtained (Figure 5a). They are used to determine the parameters of the signal I_{sa}: amplitude $A = 450$ A, phase shift between voltage and current $\varphi = 1.4$ rad. The current feedback coefficient is equal to the inverse of the nominal current. In other words, a current whose amplitude is equal to $\alpha = A \cdot k_{fbI} = 450 \cdot 1/450 = 1$ relative units, that is, the reduced value of the current amplitude is obtained. The induced current phase value is defined as $\beta = \varphi/(2 \cdot \pi) = 0.223$ relative units. In other words, the expression for the feedback current is $y = \alpha \cdot \sin(\omega \cdot t + \beta)$.

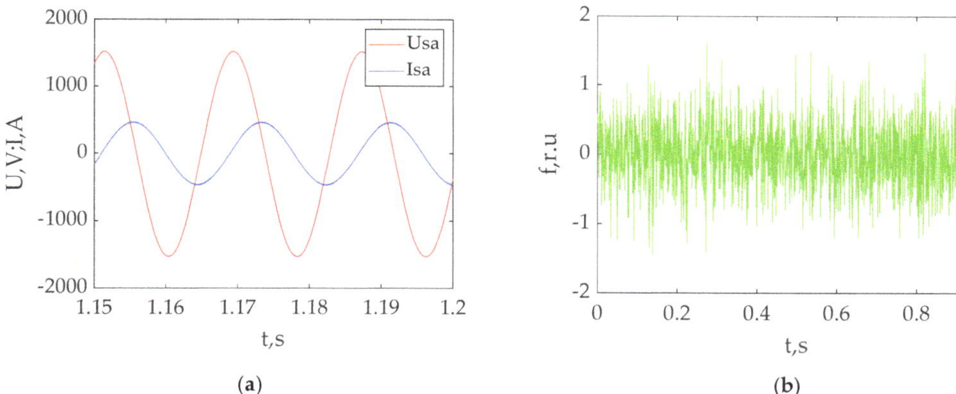

Figure 5. Time diagrams: (**a**) of the stator current and phase voltage of phase A for the steady−state mode; (**b**) 'white-noise'-type of interferences.

The 'white-noise'-type signal parameters were determined on the basis of the following considerations. The mathematical expectation of the 'white-noise'-type signal was equal to zero [46]. Lower 'white noise' dispersion was chosen than the tolerable control error equal to 5%. The chosen value of the mean squared deviation of the interference was N = 0.3375. Then, the dispersion of interference $N^2 = 0.114$. Time diagram of the interference signal is provided in Figure 5b. Since the noise is fed to the current regulator input, which receives the reduced stator current value expressed in relative units, the noise from the current sensor f is also expressed in relative units.

Spectral density of the reference signal was calculated by using formula [48]:

$$S_y(\omega) = \frac{1}{\sqrt{2 \cdot \pi}} \cdot \int_{-\infty}^{\infty} y(t) \cdot e^{-j\omega \cdot t} dt = \frac{1}{\sqrt{2 \cdot \pi}} \cdot \int_{-\infty}^{\infty} (\alpha \cdot \sin(\omega \cdot t + \beta))^2 \cdot e^{-j\omega \cdot t} dt = \frac{\alpha^2}{\omega^2 + \beta^2} \tag{9}$$

Spectral density of the interference was calculated by using formula [49]:

$$S_f(\omega) = N^2 \tag{10}$$

A hypothesis was proposed under the condition of synthesis of the optimal current controller, namely, that the signals $y(t)$ and $f(t)$ are not correlated.

3.3. Determination of the Structure and Parameters of the Optimal Current Regulator

Under the condition of synthesis of the optimal current regulator, the closed-loop system is required to perform a certain $Z(p)$ function using reference signal $Y(p)$ in accordance with transfer function $H(p)$:

$$Z(p) = H(p) \cdot Y(p) \tag{11}$$

Error of the control adjusted for interference $F(p)$:

$$e(p) = Z(p) - Y_1(p) \tag{12}$$

For this purpose, the weighting function for the closed-loop system $g^*(t)$ was determined to ensure that the mean squared error has the lowest value:

$$\bar{e}^2 = \lim_{T \to \infty} \frac{1}{T} \cdot \int_{-\infty}^{t} e^2 dt = \left| \overline{z(t) - y_1(t)} \right|^2 \to \min \tag{13}$$

where $z(t)$—the specified function and $y_1(t)$—the reference signal.

This means that the synthesized system would provide maximum suppression of the interference $f(t)$ in the reference signal. The Wiener–Hopf procedure was used to find the transfer function of the optimal controller. For this purpose, the weighting function of the system has the following expression:

$$g(t, \tau) = g^*(t, \tau) + \varepsilon \cdot \eta(t, \tau) \tag{14}$$

$\eta(t,\tau)$—some function; ε—small parameter.
Reference value $y_1(t)$ was obtained by using the convolution function:

$$y_1(t) = \int_{-\infty}^{t} [g^*(t, \tau) + \varepsilon \cdot \eta(t, \tau)] \cdot y_1(\tau) dt \tag{15}$$

Then, the filtration error is:

$$\bar{e}^2 = M\left\{ |z(t) \cdot y_1(t)|^2 \right\} = M\left\{ \left| z(t) \cdot \int_{-\infty}^{t} [g^*(t, \tau) + \varepsilon \cdot \eta(t, \tau)] \cdot y_1(\tau) d\tau \right|^2 \right\} \tag{16}$$

where $M[\cdot]$—the operation of mathematical expectation.
Optimal weighting functions were determined based on condition:

$$\frac{\partial \bar{e}^2}{\partial t} = 0 \text{ near the point } \varepsilon = 0 \tag{17}$$

$$\frac{\partial \bar{e}^2}{\partial \varepsilon} = 2 \cdot M\left\{ \left| z(t) - \int_{-\infty}^{t} g^*(t, \tau) \cdot y_1(\tau) d\tau - \int_{-\infty}^{t} \eta(t, \tau) \cdot y_1(\tau) d\tau \right| \right\} \tag{18}$$

In the integrand of the second integral (18), τ was replaced with ξ. Furthermore, the non-random character of the function $\eta(t,\xi)$ was taken into account. This enabled the authors to remove it under operator $M[\cdot]$:

$$\frac{\partial \overline{e}^2}{\partial \varepsilon} = \int\limits_{-\infty}^{t} \eta(t,\xi) \cdot \left\{ M[z(t)\cdot y_1(\xi)] - M\left[\int\limits_{-\infty}^{t} g^*(t,\tau)\cdot y_1(\xi)\cdot y_1(t)d\tau\right] \right\} d\xi \tag{19}$$

Taking into account the non-random character of $g^*(t,\tau)$ and the fact that $\eta(t,\tau) = 0$, the following was obtained:

$$M[z(t)\cdot y_1(\xi)] = \int\limits_{-\infty}^{t} g^*(t,\tau)\cdot y_1(\xi)\cdot y_1(t)d\tau \tag{20}$$

Considering the character of operation $M[\cdot]$, the Wiener–Hopf equation was obtained:

$$R_{zy}(t,\xi) = \int\limits_{-\infty}^{t} g^*(t,\tau)\cdot R_y(\tau,\xi)d\tau \tag{21}$$

where $R_{zy}(t,\xi)$—cross-correlation function of signals $z(t)$ and $y_1(t) = x(t) + f(t)$; $R_y(\tau,\xi)$—correlation function of signal $y_1(t)$; and $g^*(t,\tau)$—weighting function of the optimal closed-loop system.

Frequency transfer function of the Wiener filter was obtained from expression (21):

$$W_{opt}(j\omega) = \frac{1}{2\cdot\pi\cdot\psi(j\omega)} \cdot \int\limits_{0}^{\infty} e^{-j\omega t}dt\cdot \int\limits_{-\infty}^{\infty} \frac{S_{zy}(j\omega)}{\psi(-j\omega)}\cdot e^{j\omega t}d\omega \tag{22}$$

where:

$$S_{zy}(j\omega) = S_{zy}(j\omega) + S_{zf}(j\omega) \tag{23}$$

where $S_{zy}(j\omega)$—cross-spectral density of the wanted signal and the input signal; and $S_{zy}(j\omega)$, $S_{zf}(j\omega)$—cross-spectral densities of the signals, respectively.

The spectral density of the composite signal $y_1(t)$ was subjected to the factorization procedure:

$$S_{y1}(j\omega) = \psi(j\omega)\cdot\psi(-j\omega) \tag{24}$$

where:

$$S_{y1}(j\omega) = S_y(j\omega) + S_f(j\omega) + S_{yf}(j\omega) + S_{fy}(j\omega) \tag{25}$$

Reference signal $y(t)$ and interference $f(t)$ are not correlated to each other, so:

$$S_{yf}(j\omega) = S_{fy}(j\omega) = S_{zy}(j\omega) = 0 \tag{26}$$

Spectral density $S_{y1}(j\omega)$ is an even function of frequency ω. It can be depicted as follows:

$$S_{y1}(\omega) = \frac{b_0 + b_1\cdot\omega^2 +,\ldots,+ b_m\cdot\omega^{2m}}{a_0 + a_1\cdot\omega^2 +,\ldots,+ a_n\cdot\omega^{2n}} \tag{27}$$

where $a_0, a_1, \ldots, a_n$—coefficients at even degrees of the denominator of the function and $b_0, b_1, \ldots, b_n$—coefficients at even degrees of the numerator of the function.

From which:

$$\begin{cases} \psi(j\omega) = A\cdot\dfrac{(\omega-\gamma_1)\cdot(\omega-\gamma_2)\cdot,\ldots,\cdot(\omega-\gamma_m)}{(\omega-\lambda_1)\cdot(\omega-\lambda_2)\cdot,\ldots,\cdot(\omega-\lambda_n)}, \\[2ex] \psi(-j\omega) = A\cdot\dfrac{(\omega+\gamma_1)\cdot(\omega+\gamma_2)\cdot,\ldots,\cdot(\omega+\gamma_m)}{(\omega+\lambda_1)\cdot(\omega+\lambda_2)\cdot,\ldots,\cdot(\omega+\lambda_n)}. \end{cases} \tag{28}$$

where γ_i, λ_j—zeros and pluses of function $S_{y1}(j\omega)$:

$$A = \sqrt{\frac{b_m}{a_n}} \tag{29}$$

Condition $H(p) = 1$ was accepted. This condition corresponds to the synthesis of the optimal servo system. As a result, Equation (22) became simpler:

$$W_{opt}(j\omega) = \frac{B(j\omega)}{\psi(j\omega)} \tag{30}$$

where:

$$B(j\omega) = \sum_{i=1}^{r} \frac{a_i}{\omega - \eta_i} \tag{31}$$

where η_i—poles of function:

$$\eta_i = \frac{S_y(j\omega)}{\psi(-j\omega)} \tag{32}$$

Located in the upper half plane:

$$a_i = \left[(\omega - \eta_\iota) \cdot \frac{S_y(j\omega)}{\psi(-j\omega)} \right]_{\omega = \eta_i} \tag{33}$$

Spectral density of the output signal under the conditions of absence of correlation between reference signal $y(t)$ and interferences $f(t)$ in accordance with Equation (23) is determined as:

$$S_y(\omega) = \frac{\alpha^2}{\omega^2 + \beta^2} + N^2 \approx N^2 \cdot \frac{\omega^2 + \alpha^2}{\omega^2 + \beta^2} \tag{34}$$

Factorization of expression (34) is as follows:

$$S_y(\omega) = N \cdot \frac{\omega + j\alpha}{\omega + j\beta} \cdot N \cdot \frac{\omega - j\alpha}{\omega - j\beta} \tag{35}$$

where:

$$\psi(-j\omega) = N \cdot \frac{\omega + j\alpha}{\omega + j\beta}, \tag{36}$$

$$\psi(j\omega) = N \cdot \frac{\omega - j\alpha}{\omega - j\beta}. \tag{37}$$

whereas signals $y(t)$ and $f(t)$ are not correlated, then:

$$S_y(\omega) = \frac{\alpha^2}{\omega^2 + \beta^2} = \frac{\alpha^2}{(\omega + j\beta) \cdot (\omega - j\beta)} \tag{38}$$

In accordance with expression (32):

$$S_y(\omega) \cdot \frac{1}{\Psi(-j\omega)} = \frac{\alpha^2}{(\omega + j\beta) \cdot (\omega - j\beta)} \cdot \frac{\omega + j\beta}{N \cdot (\omega + j\alpha)} = \frac{\alpha^2}{N \cdot (\omega + j\alpha) \cdot (\omega - j\beta)} \tag{39}$$

The function has two poles: $\eta_1 = j\beta$ and $\eta_1 = -j\alpha$. Pole $\eta_1 = j\beta$ is located in the upper half plane. Then:

$$a_i = \left[(\omega - j\beta) \cdot \frac{\alpha^2}{N \cdot (\omega + j\alpha) \cdot (\omega - j\beta)} \right]_{\omega = j\beta} = \frac{\alpha^2}{jN \cdot (\beta + \alpha)} \tag{40}$$

Considering expression (31), the following was found:

$$B(j\omega) = \frac{\alpha^2}{jN\cdot(\beta+\alpha)\cdot(\omega-j\beta)} \tag{41}$$

Using Equation (30), the frequency transfer function of the optimal system was obtained:

$$W_{opt}(j\omega) = \frac{B(j\omega)}{\psi(j\omega)} = \frac{\alpha^2}{jN\cdot(\beta+\alpha)\cdot(\omega-j\beta)}\cdot\frac{(\omega-j\beta)}{N\cdot(\omega-j\alpha)} = \frac{\alpha^2}{jN^2\cdot(\beta+\alpha)\cdot(\omega-j\alpha)} = \frac{k_1}{T\cdot j\omega+1}, \tag{42}$$

where:

$$k_1 = \frac{\alpha}{N^2\cdot(\beta+\alpha)} = \frac{1}{0.3375^2\cdot(0.223+1)} = 8.752, \text{ r.u.; } T = \frac{1}{\alpha} = 1, \text{ r.u.} \tag{43}$$

Then, the structure of essentially optimal current control channel controller of the vector system for maximum suppression of the action of interference f(t) is the following:

$$W_{CRopt}(p) = \frac{W_{CRopt}(p)}{W_O(p)\cdot[1-W_{CRopt}(p)]} = \frac{k_1\cdot(1+k_{fbI}\cdot R_{el}\cdot(T_{FC}\cdot p+1)\cdot(T_1\cdot p+1))}{k_{FC}\cdot(T_1\cdot p+1-k_1)}. \tag{44}$$

Hence, expression (44) determines the structure of the optimal controller.

3.4. Simulation Results

The structural scheme of the current control channel (Figure 4) was implemented in the MATLab program environment for the basic vector control and the system of vector control with the optimal current controller. Simulation of the performance of the control scheme was performed in the presence of the reference signal y(t) and interference f(t) at the controller input. The time diagrams of the stator current of phase A of the traction motor were obtained for the basic control scheme and for the control scheme with the optimal controller (Figure 6).

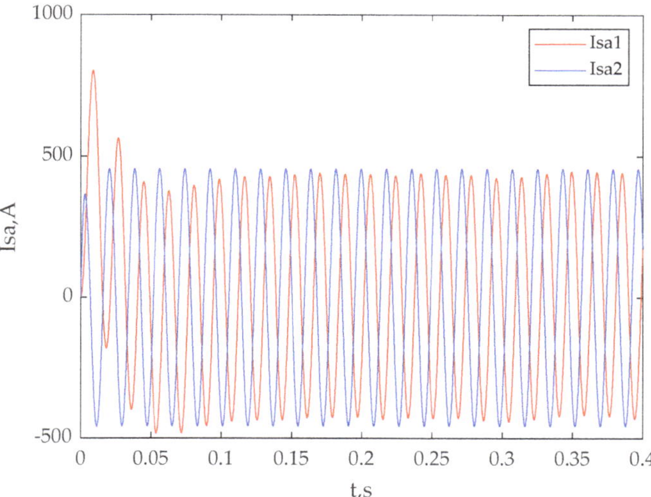

Figure 6. Time diagrams of stator current of phase A of the traction motor: I_{sa1}—of the basic scheme of vector control; I_{sa2}—of the control system with the optimal controller.

As suggested in Figure 6, the system of vector control with the optimal current controller is more resistant to the action of interferences.

To compare the quality indicators of control of both the systems, the transient characteristics of the basic system of vector control and of the system with the optimal controller were obtained. For this purpose, a unit step-type signal was fed to the input of the both systems (45) under the condition of absence of the interference current at the controller input:

$$x(t) = \begin{cases} 0, & \text{if } t < 0, \\ 1, & \text{if } t \geq 0. \end{cases} \tag{45}$$

The transient functions of the basic vector control system (I_{sa1}) and the vector system with the optimal controller (I_{sa2}) are presented in Figure 7a. The time values of the transient process for the basic system (t_{st1}) and the system with optimal current controller (t_{st2}), the steady-state value of the stator current (I_{sa1st}) and (I_{sa2st}), and the maximum value of the stator current (I_{sa1max}) and (I_{sa2max}), respectively, were determined based on the transient characteristics. The results are presented in Table 3. The overshooting of both systems was calculated using the formula:

$$\sigma = \frac{I_{samax} - I_{sast}}{I_{sast}} \cdot 100\%. \tag{46}$$

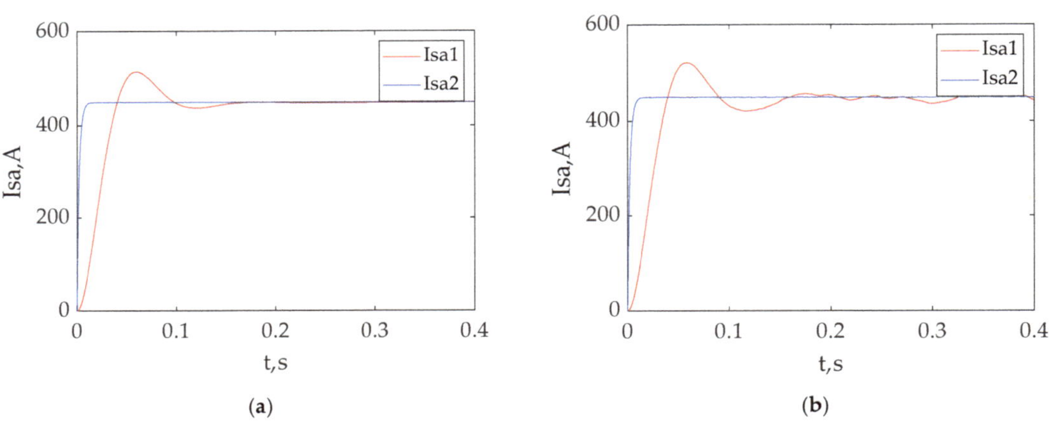

(a) (b)

Figure 7. Transient functions: (**a**) under the absence of an interference at the current controller input (of the basic control system (I_{sa1}) and system with the optimal current controller (I_{sa2st})); (**b**) under the presence of an interference at the current controller input (of the basic control system (I_{sa1}) and system with the optimal current controller (I_{sa2st})).

Table 3. The results of comparison of the control quality of the basic system versus the system with optimal controller.

Parameter	Designation	Unit of Measurement	Value	
			Basic System	System with Optimal Controller
Time of transient process	t_{st}	s	0.166	0.0138
Steady-state values	$I_{sa.st}$	A	450	450
Maximum value	$I_{sa.max}$	A	514.4	450
Overshooting	σ	%	14.3	0
Control accuracy	ε	%	1.3151	0.0134

The results of the overshoot calculations are presented in Table 3.

To determine the precision of control of both control systems, transient functions were obtained under the condition of the presence of the 'white-noise'-type of interference controllers at the input. Transient functions under the condition of interferences of the basic control system (I_{sa1}) and the system with the optimal current controller (I_{sa2st}) are presented in Figure 7b.

The control errors were calculated for the steady-state mode of both systems using formula:

$$\varepsilon = \frac{\frac{1}{N-1} \cdot \sum_{i=0}^{N} \left(I_{sast} - I_{sai} \right)^2}{I_{sast}} \cdot 100\% \tag{47}$$

where: I_{sast}—steady-state value of stator current; N—number of samples.

The results of calculations are presented in Table 3.

The control quality parameter, such as the time of the oscillatory process, was not taken into account due to the lack of its relevance in the investigation. Since there is no overshoot in the proposed regulator as opposed to the basic regulator (Figure 7), this means that the current control channel with the proposed regulator has an infinite phase stability margin.

4. Discussion

In the present study, the deterministic signal was used as the reference signal and the stochastic signal was used as the interference.

This circumstance can be explained by the following two factors:

1. Objective difficulties in obtaining the experimental data on the dependence of the stator current of the traction motor in the case of the stochastic character of the change of the phase voltage;
2. Performance of the existing models of the asynchronous motor under the stochastic character of change of the phase voltage was not analyzed. The results of the simulation of performance of the asynchronous traction drive with the vector control system may be flawed.

The second factor also explains the choice of only the current control channel of the vector control system for the investigation.

The quality of the current control channel of the vector control system for the asynchronous traction drive was determined according to its transient characteristics under the condition of absence of interferences at the input (Figure 7a). Comparison of the quality indicators of the basic system and of the system with the optimal controller (Table 3) showed that:

— The steady-state value of the output signal of the both systems was equal to 450 A. This value corresponded to the value of the nominal phase current of the asynchronous traction motor;
— The speed of response of the system with the optimal controller was higher than that of the basic system. Time of the transient process (Table 3) of the system with optimal controller $t_{st1} = 0.0138$ s, of the basic system—$t_{st1} = 0.166$ s;
— Transient function of the system with the optimal controller was free from oscillations. The basic system had oscillations. Value of overshooting of the basic system was 14.3%.

The accuracy of the overshooting of the current control channels was determined based on the transient functions of the systems in the presence of 'white-noise'-type interference at the controller input (Figure 7b). The comparison of simulation results (Table 3) showed that the system with the optimal controller had a 1.3% lower control error than the basic system. This fact suggests that the power loss caused by control interferences is 1.69% lower in the traction drive with the optimal controller than in the drive with the basic control system. This can be supported by the phase current diagrams of the stator (Figure 6) under the presence of interferences at the controller input. For the steady-state mode, the phase current value of the stator of the traction motor with the basic control system was 444 A, with the optimal controller—450 A.

There are some important caveats to the study that deserve mention:

1. The current control channel models of the control circuits were based on the assumption that the control signal at the input of the current controllers is a deterministic signal;
2. The character of load variation was not taken into account (it can have a stochastic nature);
3. The investigations were carried out for the mode of operation, when the value of the shaft rotation speed is equal to the value of the given rotation speed;
4. The optimal current controller will work correctly for the set operating mode of the traction drive system. This fact is related to the condition that the parameters of the optimal current regulator depend on the values of stator current and noise variance. During transients, these parameters change.

Nonetheless, it should be taken into account that the models of the current control channels were based on the assumption that the reference signal at the current controller input was a deterministic signal. This factor implies certain limitations to the application of the model developed. To account for this factor, additional investigations must be conducted. At the same time, the authors realize the difficulties related to retrieval of the experimental data under the conditions of operation of an electric locomotive.

In the process of work on the present paper, the authors encountered objective difficulties related to the absence of any possibility to conduct a full-scale experiment and retrieve valid experimental data. This is related to the fact that the control systems are essentially microprocessor systems by design. To record the experimental oscillograms of the signals at the current control input, it is necessary to obtain permission from the respective manufacturers of the systems. This is difficult to implement as this is related with trade secrets.

Further work to improve on these new developments is suggested:

1. Investigation of the influence of the stochastic nature of voltage changes in the contact network on the supply voltage of the autonomous voltage inverter;
2. Investigation of the influence of thermal noise of the autonomous voltage inverter on the operation of the vector control system;
3. Investigation of the thermal noise of traction motor windings on the operation of the vector control system;
4. Investigation of the stochastic nature of the traction drive load on the operation of the vector control system;
5. Development of a current regulator, whose parameters are adapted to the operating conditions of the locomotive.

5. Conclusions

1. The structural scheme was obtained in the transfer functions of the current control channel of the vector control systems for the asynchronous traction drives in the AC electric locomotives. Spectral densities of the reference signal and of the interference at the current controller input were calculated.
2. The design synthesis was performed, and parameters of optimal current controller parameters were obtained by using the Wiener–Hopf equation for the vector control system for the traction drive in the electric locomotive with the asynchronous motor.
3. Simulation of the current control channel of the basic vector control system and of the system with the optimal controller was performed.
4. The transient characteristics of the current control channels of the basic control system and the optimal controller were obtained under the condition of absence of interferences at the controller input. Comparison of the time of the transient processes showed that:
   - The speed of response of the system with the optimal controller was higher by 1.3 s compared to the basic control system;
   - Transient characteristic of the system with the optimal controller was free from oscillations. The oscillations were present in the transient characteristic of the basic system. The overshoot of the basic system was equal to 14.3%.
   - The steady-state value of the transient function of the both systems was 450 A, which corresponded to the nominal value of the phase current of the stator of the traction motor.

The accuracies of control of the two systems were calculated on the basis of the transient characteristics of the current control channels under the condition of the presence of 'white noise' interferences at the controller input. The accuracy of the control of the system with the optimal controller was higher by 1.3% than that of the basic system.

Analysis of the time diagrams of the phase currents of the traction drive showed that the presence of interferences at the input of the current controller in the optimal controller did not influence the value of the phase current of the motor stator. In the basic system, the value of the phase current of the stator was lower by 1.3% compared to the operation without interferences. This fact suggests that the power loss is 1.69% lower in the traction drive with the optimal controller than in the traction drive with the basic control system.

6. Patents

Implementation of the present study was financially supported by the Ministry of Education and Science of Ukraine under the framework of the R&D project 'Improvement of energy efficiency of the railway rolling stocks on the basis of the resource-saving technologies and intelligent power systems' (state registration number 0120U101912).

Author Contributions: Conceptualization, methodology, and investigation S.G., V.T., S.S. and B.L.; investigation, writing—review and editing, S.G., V.L. and R.K.; formal analysis, writing—review and editing, R.M. and A.K.; investigation, S.G., V.T., S.S. and B.L. All authors have read and agreed to the published version of the manuscript.

Funding: This publication was realized with the support of Operational Program Integrated Infrastructure 2014–2020 of the project: Innovative Solutions for Propulsion, Power, and Safety Components of Transport Vehicles, code ITMS 313011V334, co-financed by the European Regional Development Fund.

Institutional Review Board Statement: Not applicable.

Informed Consent Statement: Not applicable.

Data Availability Statement: Not applicable.

Conflicts of Interest: The authors declare no conflict of interest.

References

1. Kostin, M.; Mishchenko, T.; Hoholyuk, O. Fryze Reactive Power in Electric Transport Systems with Stochastic Voltages and Currents. In Proceedings of the 2020 IEEE 21st International Conference on Computational Problems of Electrical Engineering (CPEE), Online, 16–19 September 2020; pp. 1–4. [CrossRef]
2. Feng, D.; Yang, C.; Cui, Z.; Li, N.; Sun, X.; Lin, S. Research on Optimal Nonperiodic Inspection Strategy for Traction Power Supply Equipment of Urban Rail Transit Considering the Influence of Traction Impact Load. *IEEE Trans. Transp. Electrif.* **2020**, *6*, 1312–1325. [CrossRef]
3. Goolak, S.; Tkachenko, V.; Bureika, G.; Vaičiūnas, G. Method of Spectral Analysis of Traction Current of AC Electric Locomotives. *Transport* **2021**, *35*, 658–668. [CrossRef]
4. Kostin, M.; Nikitenko, A.; Mishchenko, T.; Shumikhina, L. Electrodynamics of Reactive Power in the Space of Inter-Substation Zones of AC Electrified Railway Line. *Energies* **2021**, *14*, 3510. [CrossRef]
5. Morris, J.; Robinson, M.; Palacin, R. Use of Dynamic Analysis to Investigate the Behaviour of Short Neutral Sections in the Overhead Line Electrification. *Infrastructures* **2021**, *6*, 62. [CrossRef]
6. Mariscotti, A. Critical Review of EMC Standards for the Measurement of Radiated Electromagnetic Emissions from Transit Line and Rolling Stock. *Energies* **2021**, *14*, 759. [CrossRef]
7. Song, Y.; Wang, H.; Liu, Z. An Investigation on the Current Collection Quality of Railway Pantograph-Catenary Systems with Contact Wire Wear Degradations. *IEEE Trans. Instrum. Meas.* **2021**, *70*, 1–11. [CrossRef]
8. Mariscotti, A. Data sets of measured pantograph voltage and current of European AC railways. *Data Brief* **2020**, *30*, 105477. [CrossRef]
9. Song, Y.; Antunes, P.; Pombo, J.; Liu, Z. A methodology to study high-speed pantograph-catenary interaction with realistic contact wire irregularities. *Mech. Mach. Theory* **2020**, *152*, 103940. [CrossRef]
10. Mikhailov, E.; Sapronova, S.; Tkachenko, V.; Semenov, S.; Smyrnova, I.; Kholostenko, Y. Improved solution of guiding of railway vehicle in curves. In Proceedings of the 23th International Scientific Conference "Transport Means", Palanga, Lithuania, 2–4 October 2019; pp. 916–921. Available online: https://transportmeans.ktu.edu/wp-content/uploads/sites/307/2018/02/Transport-means-2019-Part-2.pdf (accessed on 20 December 2021).

11. Grebennikov, N.; Kharchenko, P. Development of a Computer Model of a Passenger Train Using Data from Devices for Train Operation Parameters Registration. In Proceedings of the 2021 International Conference on Industrial Engineering, Applications and Manufacturing (ICIEAM), Sochi, Russia, 17–21 May 2021; pp. 908–913. [CrossRef]
12. Babyak, M.; Keršys, R.; Neduzha, L. Improving the Dependability Evaluation Technique of a Transport Vehicle. In Proceedings of the 24th International Scientific Conference "Transport Means", Palanga, Lithuania, 30 September–2 October 2020; pp. 646–651. Available online: http://eadnurt.diit.edu.ua/jspui/handle/123456789/12287 (accessed on 20 December 2021).
13. Zhou, Z.; Chen, Z.; Spiryagin, M.; Wolfs, P.; Wu, Q.; Zhai, W.; Cole, C. Dynamic performance of locomotive electric drive system under excitation from gear transmission and wheel-rail interaction. *Veh. Syst. Dyn.* **2021**, 1–23. Available online: https://www.tandfonline.com/doi/full/10.1080/00423114.2021.1876887 (accessed on 20 December 2021). [CrossRef]
14. Goolak, S.; Sapronova, S.; Tkachenko, V.; Riabov, I.; Batrak, Y. Improvement of the model of power losses in the pulsed current traction motor in an electric locomotive. *East. Eur. J. Enterp. Technol.* **2020**, *6*, 38–46. [CrossRef]
15. Rouamel, M.; Gherbi, S.; Bourahala, F. Robust stability and stabilization of networked control systems with stochastic time-varying network-induced delays. *Trans. Inst. Meas. Control* **2020**, *42*, 1782–1796. [CrossRef]
16. Wang, B.; Zhu, Q. Stability Analysis of Discrete-Time Semi-Markov Jump Linear Systems. *IEEE Trans. Autom. Control* **2020**, *65*, 5415–5421. [CrossRef]
17. Jiao, T.; Zong, G.; Pang, G.; Zhang, H.; Jiang, J. Admissibility analysis of stochastic singular systems with Poisson switching. *Appl. Math. Comput.* **2020**, *386*, 125508. [CrossRef]
18. Ronanki, D.; Singh, S.A.; Williamson, S.S. Comprehensive Topological Overview of Rolling Stock Architectures and Recent Trends in Electric Railway Traction Systems. *IEEE Trans. Transp. Electrif.* **2017**, *3*, 724–738. [CrossRef]
19. Pugachev, A. Efficiency increasing of induction motor scalar control systems. In Proceedings of the 2017 International Conference on Industrial Engineering, Applications and Manufacturing (ICIEAM), St. Petersburg, Russia, 16–19 May 2017; pp. 1–5. [CrossRef]
20. Costa, C.A.; Nied, A.; Nogueira, F.G.; Turqueti, M.D.A.; Rossa, A.J.; Dezuo, T.J.M.; Barra, W. Robust Linear Parameter Varying Scalar Control Applied in High Performance Induction Motor Drives. *IEEE Trans. Ind. Electron.* **2020**, *68*, 10558–10568. [CrossRef]
21. Novak, H.; Lesic, V.; Vasak, M. Energy-Efficient Model Predictive Train Traction Control with Incorporated Traction System Efficiency. *IEEE Trans. Intell. Transp. Syst.* **2021**, 1–12. Available online: https://ieeexplore.ieee.org/document/9325928 (accessed on 22 December 2021). [CrossRef]
22. Jumaev, O.A.; Sayfulin, R.R.; Samadov, A.R.; Arziyev, E.I.; Jumaboyev, E.O. Digital control systems for asynchronous electrical drives with vector control principle. *IOP Conf. Ser. Mater. Sci. Eng.* **2020**, *862*, 032054. [CrossRef]
23. Botirov, T.V.; Latipov, S.B.; Buranov, B.M. About one synthesis method for adaptive control systems with reference models. *J. Phys. Conf. Ser.* **2020**, *1515*, 022078. [CrossRef]
24. Lee, J.-K.; Kim, J.-W.; Park, B.-G. Fast Anti-Slip Traction Control for Electric Vehicles Based on Direct Torque Control with Load Torque Observer of Traction Motor. In Proceedings of the 2021 IEEE Transportation Electrification Conference & Expo (ITEC), Chicago, IL, USA, 21–25 June 2021; pp. 321–326. [CrossRef]
25. Aissa, B.; Hamza, T.; Yacine, G.; Mohamed, N. Impact of sensorless neural direct torque control in a fuel cell traction system. *Int. J. Electr. Comput. Eng. (IJECE)* **2021**, *11*, 2725–2732. [CrossRef]
26. Karlovsky, P.; Lettl, J. Induction Motor Drive Direct Torque Control and Predictive Torque Control Comparison Based on Switching Pattern Analysis. *Energies* **2018**, *11*, 1793. [CrossRef]
27. Kumar, Y.S.; Poddar, G. Medium-Voltage Vector Control Induction Motor Drive at Zero Frequency Using Modular Multilevel Converter. *IEEE Trans. Ind. Electron.* **2017**, *65*, 125–132. [CrossRef]
28. Hassan, M.M.; Shaikh, M.S.; Jadoon, H.U.K.; Atif, M.R.; Sardar, M.U. Dynamic Modeling and Vector Control of AC Induction Traction Motor in China Railway. *Sukkur IBA J. Emerg. Technol.* **2020**, *3*, 115–125. [CrossRef]
29. Wang, H.; Liu, Y.; Ge, X. Sliding-mode observer-based speed-sensorless vector control of linear induction motor with a parallel secondary resistance online identification. *IET Electr. Power Appl.* **2018**, *12*, 1215–1224. [CrossRef]
30. Ćwil, M.; Bartnik, W.; Jarzębowski, S. Railway Vehicle Energy Efficiency as a Key Factor in Creating Sustainable Transportation Systems. *Energies* **2021**, *14*, 5211. [CrossRef]
31. Nurali, P.; Mirzaev, U. Ways to Improve the Energy Efficiency of an Electric Drive with Asynchronous Mo-tors. *Int. J. Eng. Inf. Syst. (IJEAIS)* **2021**, *5*, 230–233.
32. Pleshivtseva, Y.E.; Rapoport, E.Y. Parametric Optimization of Systems with Distributed Parameters in Problems with Mixed Constraints on the Final States of the Object of Control. *J. Comput. Syst. Sci. Int.* **2018**, *57*, 723–737. [CrossRef]
33. Pleshivtseva, Y.E.; Rapoport, E.Y. Optimal Energy-Efficient Programmed Control of Distributed Parameter Systems. *J. Comput. Syst. Sci. Int.* **2020**, *59*, 518–532. [CrossRef]
34. Rapoport, E.Y. Method for Parametric Optimization in Problems of the Multichannel Control of Systems with Distributed Parameters. *J. Comput. Syst. Sci. Int.* **2019**, *58*, 545–559. [CrossRef]
35. Razghonov, S.; Kuznetsov, V.; Zvonarova, O.; Chernikov, D. Track circuits adjusting calculation method under current influence traction interference and electromagnetic compatibility. *IOP Conf. Ser. Mater. Sci. Eng.* **2020**, *985*, 012017. [CrossRef]
36. Çelik, E. Incorporation of stochastic fractal search algorithm into efficient design of PID controller for an automatic voltage regulator system. *Neural Comput. Appl.* **2018**, *30*, 1991–2002. [CrossRef]

37. Weerakkody, S.; Liu, X.; Sinopoli, B. Robust structural analysis and design of distributed control systems to prevent zero dynamics attacks. In Proceedings of the 2017 IEEE 56th Annual Conference on Decision and Control (CDC), Melbourne, Australia, 12–15 December 2017; pp. 1356–1361. [CrossRef]
38. Kiumarsi, B.; Vamvoudakis, K.G.; Modares, H.; Lewis, F.L. Optimal and Autonomous Control Using Reinforcement Learning: A Survey. *IEEE Trans. Neural Netw. Learn. Syst.* **2018**, *29*, 2042–2062. [CrossRef] [PubMed]
39. Luo, B.; Yang, Y.; Liu, D.; Wu, H.-N. Event-Triggered Optimal Control with Performance Guarantees Using Adaptive Dynamic Programming. *IEEE Trans. Neural Netw. Learn. Syst.* **2020**, *31*, 76–88. [CrossRef]
40. Possieri, C.; Sassano, M. Deterministic Optimality of the Steady-State Behavior of the Kalman–Bucy Filter. *IEEE Control Syst. Lett.* **2019**, *3*, 793–798. [CrossRef]
41. Feliu-Batlle, V.; Feliu-Talegón, D.; San-Millan, A.; Rivas-Pérez, R. Wiener-Hopf optimal control of a hydraulic canal prototype with fractional order dynamics. *ISA Trans.* **2018**, *82*, 130–144. [CrossRef]
42. Zhao, G.; Qian, X.; Yoon, B.-J.; Alexander, F.J.; Dougherty, E.R. Model-Based Robust Filtering and Experimental Design for Stochastic Differential Equation Systems. *IEEE Trans. Signal Process.* **2020**, *68*, 3849–3859. [CrossRef]
43. Goolak, S.; Liubarskyi, B.; Sapronova, S.; Tkachenko, V.; Riabov, I.; Glebova, M. Improving a model of the induction traction motor operation involving non-symmetric stator windings. *East. Eur. J. Enterp. Technol.* **2021**, *4*, 45–58. [CrossRef]
44. Goolak, S.; Tkachenko, V.; Šťastniak, P.; Sapronova, S.; Liubarskyi, B. Analysis of Control Methods for the Traction Drive of an Alternating Current Electric Locomotive. *Symmetry* **2022**, *14*, 150. [CrossRef]
45. Yousefi-Talouki, A.; Pescetto, P.; Pellegrino, G.-M.L.; Boldea, I. Combined Active Flux and High-Frequency Injection Methods for Sensorless Direct-Flux Vector Control of Synchronous Reluctance Machines. *IEEE Trans. Power Electron.* **2018**, *33*, 2447–2457. [CrossRef]
46. Al-Gabalawy, M.; Hosny, N.S.; Dawson, J.A.; Omar, A.I. State of charge estimation of a Li-ion battery based on extended Kalman filtering and sensor bias. *Int. J. Energy Res.* **2021**, *45*, 6708–6726. [CrossRef]
47. Goolak, S.; Liubarskyi, B.; Sapronova, S.; Tkachenko, V.; Riabov, I. Refined Model of Asynchronous Traction Electric Motor of Electric Locomotive. In Proceedings of the 25th International Scientific Conference "Transport Means", Online, 6–8 October 2021; pp. 455–460.
48. Valluri, S.R.; Dergachev, V.; Zhang, X.; Chishtie, F.A. Fourier transform of the continuous gravitational wave signal. *Phys. Rev. D* **2021**, *104*, 024065. [CrossRef]
49. Han, X.; Xue, L.; Shao, F.; Xu, Y. A Power Spectrum Maps Estimation Algorithm Based on Generative Adversarial Networks for Underlay Cognitive Radio Networks. *Sensors* **2020**, *20*, 311. [CrossRef] [PubMed]

 *energies*

Article

Dynamic Impact of a Rail Vehicle on a Rail Infrastructure with Particular Focus on the Phenomenon of Threshold Effect

Włodzimierz Idczak *, Tomasz Lewandrowski, Dominik Pokropski, Tomasz Rudnicki and Jacek Trzmiel

Faculty of Civil Engineering and Geodesy, Military University of Technology, 2 Gen. Sylwestra Kaliskiego Str., 00-908 Warsaw, Poland; tomasz.lewandrowski@wat.edu.pl (T.L.); dominik.pokropski@wat.edu.pl (D.P.); tomasz.rudnicki@wat.edu.pl (T.R.); jacek.trzmiel@wat.edu.pl (J.T.)
* Correspondence: wlodzimierz.idczak@wat.edu.pl

Abstract: The paper undertakes analysis of the dynamic impact of a rail vehicle on various types of a railway infrastructure with particular focus on the phenomenon of threshold effect within the transition zones of an engineering facility. The problem of locally variable stiffness of the railway infrastructure, which in turn could lead to the accelerated infrastructure degradation, is identified. Using the analytical and numerical background, the computational model is presented, based on which, it could be possible to determine the impact of the various rail support on the dynamic response of the entire infrastructure. The dynamic load, caused by the passage of the multiaxle rail vehicle, is taken into account in the paper. The fourth-order differential equation is solved by using the finite differences method with application of the numerical MATLAB script. The created numerical algorithm and a number of calculations allowed the formulation of several solutions that could reduce the dynamic impact of the rail vehicle on the railway surface within the transition zones. In the paper, theoretical results are compared to the field measurements conducted on a real dynamically loaded rail. Field experiments have been carried out on the railway track in operation. The vertical displacement of a rail, dynamically loaded by various types of rail vehicles passing by (both passenger and freight trains) has been investigated. Researches have been carried out in the area of transition zones of engineering facilities. Test points have been selected in places where there is a sudden change in parameters of the track structure (e.g., a change from concrete sleepers to wooden sleepers). Based on conducted researches it has been possible to validate results obtained from the numerical calculations.

Keywords: rail vehicle; rail infrastructure; threshold effect; dynamics; numerical simulations; testing

Citation: Idczak, W.; Lewandrowski, T.; Pokropski, D.; Rudnicki, T.; Trzmiel, J. Dynamic Impact of a Rail Vehicle on a Rail Infrastructure with Particular Focus on the Phenomenon of Threshold Effect. *Energies* **2022**, *15*, 2119. https://doi.org/10.3390/en15062119

Academic Editors: Larysa Neduzha and Jan Kalivoda

Received: 15 February 2022
Accepted: 9 March 2022
Published: 14 March 2022

Publisher's Note: MDPI stays neutral with regard to jurisdictional claims in published maps and institutional affiliations.

Copyright: © 2022 by the authors. Licensee MDPI, Basel, Switzerland. This article is an open access article distributed under the terms and conditions of the Creative Commons Attribution (CC BY) license (https://creativecommons.org/licenses/by/4.0/).

1. Introduction

The problem of locally variable stiffness of the railway surface within the transition zones of an engineering facility has been identified, which in turn may lead to accelerated degradation of the structure and the need to incur increased expenditures on maintaining the infrastructure in proper condition.

The aim of the research and computational work was to analyze the dynamic impact of the rail vehicle on various solutions of the railway surface structure, with particular emphasis on the phenomenon of the threshold effect within the transition zones of the engineering facility.

The basic task of the surface is to enable safe and stable driving of a rail vehicle on a specific trajectory and to take over loads from the wheels of a rail vehicle and transfer them to the subtrack. Two basic types of surface can be distinguished: (1) classic surface (ballast) and (2) unconventional surface (ballastless).

In ballastless surfaces, rubble has been replaced with layers of materials with different modulus of elasticity. They are arranged in such a way that materials with lower modules are built into the lower layers of the structure and the higher layers have the higher modulus of elasticity. In this way, each subsequent layer of the structure (sub-track, frost-resistant layer, asphalt-stabilized layer/hydraulically stabilized layer, concrete support layer, primer, rail) has increasing rigidity. The thicknesses and materials of individual layers should be selected so that the structure works in the field of elastic deformation [1]. Constructions of ballastless surfaces can be divided as follows: (1) surfaces in which the supporting layer consists of prefabricated slabs (IPA, Bogl, VA Shinkansen, OBB-PORR), (2) surfaces in which the supporting layer is a concrete slab (Rheda, EBS System) or bituminous layer (Getrac) laid directly on the construction site and (3) surfaces in which the rail is located in specially prepared rail channels (ERS, Infudo system, BBEST system) [2–7].

Due to a number of advantages of ballastless surfaces (including lower thickness and lower weight), they are often used in engineering structures such as bridges, viaducts, as well as in tunnels. If there is a ballast surface on the trail and on the engineering object a ballastless surface, one may observe the so-called threshold effect on the access to the facility and behind it.

The phenomenon of the threshold effect is due to the different parameters of adjacent different surfaces. This is especially noticeable in the vicinity of engineering facilities. In places of changing the type of surface, in the zones in front of and behind the object, vertical irregularities of rails (basins) are formed, which increase during operation and cause further deformations. As a result of dynamic loads, the track grate often rises, the track twist increases, as well as uneven wear of the rails and damage to the fasteners on both types of surfaces. Gaps may form under the sleepers, which threatens the stability of the structure. The threshold effect has a negative impact not only on the railway surface, but also on the object that is exposed to excessive loads and vibrations.

The causes of the threshold effect occurrence can be divided into primary (mechanical) and secondary (geometric). Among the primary causes, the following stand out: a change in the elasticity of the ground substrate, a change in bending stiffness of the supporting system, a change in the mass of the surface and a change in the surface damping value. The secondary causes include: the constantly occurring primary effect, the settling of a backstrip and a subtrack as a result of dynamic loads and vibrations and the unevenness and geometric defects of the rails [8].

The threshold effect causes an increase in the wheel-rail interactions and overloads of the bottom and subtrack. In the zones in front of and behind the object, vertical irregularities of rails (basins) are formed, which increase during operation and cause further deformations. As a result of dynamic loads, the track grate often rises, the track twist increases, as well as uneven wear of the rails and damage to fasteners on both types of surfaces. Gaps may form under the sleepers, which threatens the stability of the structure.

A significant impact on the threshold effect is the phenomenon of different effective stiffness of rail bending in the case of ballast and ballastless surfaces. Fastened with fastenings, the rail is subject to cyclic pressing and lifting. In the case of ballastless surfaces, the force needed to raise the rail is much greater, thanks to which there is no local loss of mutual contact between individual elements of the surface. The essence of greater effective bending rigidity is particularly visible: (1) in the case of the rail systems in the sheath, where the rail is embedded in an elastic mass that limits its freedom of bending and (2) in the case of direct attachment of the rail to the support plate or fastening the rail to bridges [9].

The threshold effect has a negative impact not only on the railway surface, but on the object that is exposed to excessive loads and vibrations, as well [8–10]. It is assumed that the difference in track stiffness on and off the object should not be more than 30%. The transition from a ballast to a ballastless surface is shown in Figure 1.

In the course of the topic implementation, for the purpose of performing this analysis of the dynamic impact of a vehicle on the railway surface, a computational model was created, allowing determination of the impact of differentiated rail support on the dynamic response of the entire structure. The starting point for the considerations was the Bernoulli-Euler beam, located on the elastic Winkler substrate. The dynamic load, caused by the passage of a multiaxle rail vehicle and the different parameters of different types of surfaces, were taken into account. As a consequence, a fourth-order differential equation was obtained. It was solved by the finite differences method. A script in MATLAB was developed for a numerical solution of the problem. At the same time, it should be emphasized that development of an algorithm using the finite differences method does not involve the need to use complicated and expensive computer software.

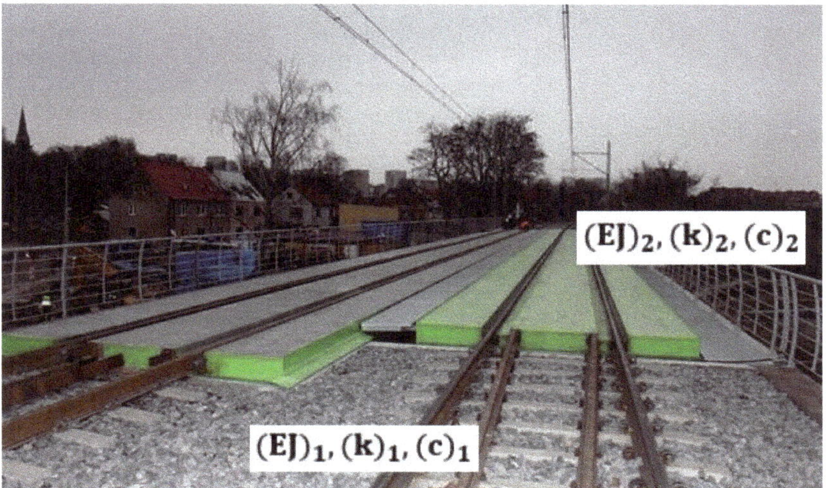

Figure 1. Transition from a ballast to a ballastless rail surface as a cause of the threshold effect.

In order to verify and validate the created algorithm, in situ studies of vertical displacement of a dynamically loaded railway rail were carried out. The research was carried out using laser scanning technology.

2. Materials and Methods

2.1. Materials

The railway surface is affected by technical and operational factors. Among the technical factors, one should mention such parameters of the structure itself as: the type of rail steel used and its longitudinal elastic modulus; the type of rail and the geometric moment of inertia of its cross-section and the type of surface; vibration-damping coefficient; type of sleepers, their spacing; the subtrack structure, and the resulting modulus of elasticity of the substrate. Operational factors include the type of rolling stock and its chassis layout, the wheelbase of the vehicle and the load on its single axle, as well as the speed of passage on the surface.

In structural calculations of the railway surface, an important parameter is the longitudinal elastic modulus (or Young's modulus). For the rail steel, its value is 210 GPa [11].

The study assumes a geometric moment of inertia of the cross-section of the 60E1 rail equal to 3038.3 cm^4 with a mass of 60.21 kg/m.

In the literature of the subject for calculations, a model of the railway surface is adopted, in which the rail is based on the elastic Winkler substrate characterized by the modulus of elasticity of the substrate. The validity of this assumption has been confirmed by previous research and analytical work—among others, in paper [12]—the sufficiency of the elastic substrate model for the analysis of such issues was confirmed. The value of the elastic modulus of the substrate is influenced by the materials from which the subsequent layers of the surface were made, and their thickness. These are a rail washer, sleeper, ballast, subtrack, a concrete or asphalt slab (instead of a ballast) in the case of unconventional surfaces, and, in the case of a surface on an engineering object, the structure of the object itself (instead of a subtrack). The value of the elastic modulus of the entire surface "k" can be calculated from the following relationship [13]:

$$\frac{1}{k} = \sum_i \frac{1}{k_i} \, ,$$

(1)

where: k—modulus of elasticity of the entire surface, k_i—modulus of elasticity of layer "i".

The elastic modulus of the rail washer is at the level of (90–100) MPa [14], while the value of the modulus of elasticity of the prestressed concrete primer is at the level of 31 GPa and the value of the wooden primer is at the level of (9.4–10) GPa [15]. For the concrete supporting layer, it is about 34 GPa; for the asphalt-stabilized layer, 5 GPa; and for the hydraulically stabilized layer, 10 GPa [1]. For the ballast layer, the modulus of elasticity is (250–300) MPa [15], and for the subtrack (40–120) MPa [16,17]. For a reinforced concrete engineering object, the modulus of elasticity is at the level of 28.5 GPa [17]. Taking into account the above values and relation (1), as well as using the data provided in [18,19], elastic moduli for various structures of the railway surface were determined. The results are presented in Table 1.

Table 1. Modulus of elasticity for different types of railway surface.

Surface Type	Elastic Modulus Value [MPa]
ballast surface (wooden sleepers)	24–35
ballast surface (concrete sleepers)	42–46
ballast surface with frozen ballast layer	70–90
ballastless surface	92–102

Precisely determining the value of the elastic modulus of the substrate is difficult. Its size can be influenced by many factors, such as: ambient temperature, soil humidity, infrastructure maintenance status, surface age, or the current transferred load [19].

With the passage of time after applying the load, the amplitude of the system vibrations decreases its value. This is due to the phenomenon of vibration damping. The vibration-damping force is an action inside the structure that opposes the load. The following are distinguished in the constructions: (1) structural damping and (2) material damping. Structural damping is caused by the connection and cooperation of individual elements of the structure. Material attenuation is caused by the structure of the material and internal friction [20]. The value of the damping force C is described by the relation:

$$C = c \cdot \frac{dw}{dt} \, ,$$

(2)

where: C—damping force, c—vibration-damping coefficient, dw/dt—change of vertical displacement of the rail as a function of time.

Based on the literature for the ballast railway surface, a vibration-damping coefficient of 22.6 MNs/m^2 [17] was assumed. The analysis of the literature shows that ballastless surfaces are characterized by worse damping properties due to greater rigidity. For the purposes of calculations, a vibration-damping coefficient was assumed for this type of construction with a value 15% lower than for the classic surfaces.

The chassis system determines the way in which the dynamic loads generated by the rail vehicle are transmitted to the surface. The load is transmitted pointwise, in the places of contact of the wheel with the rail. The static scheme of this system is a series of concentrated forces, applied to the rail in the spacing defined by the chassis design. The chassis of locomotives, as a rule, consists of two trolleys in a system of two or three axles each. The wheelbase in the trolley is (2.60–4.15) m. The wheelbase of the wagon bogie is equal from 1.5 m to even 8.0 m.

The axle load of a rail vehicle must not exceed the limit values specified by the Railway Infrastructure Manager. Depending on the type and condition of the surface, these values are determined individually for each section of the railway line. In Poland, on railway lines managed by Polskie Linie Kolejowe, the axle load of a rail vehicle may not exceed 221 kN/axle [21].

Currently, the world record for instantaneous speed developed by a conventional rail vehicle is 574.8 km/h. It was established by the French TGV V150 train on 3 April 2007 on the TGV Est line between Strasbourg and Paris [22]. However, commercial travel speeds are much lower. For shunting driving, this can be a speed of even less than 10 km/h, and for high-speed passenger trains (200–300) km/h. The higher the speed of rail vehicles, the higher the requirements for proper diagnostics and maintenance of the surface are greater and more restrictive. Driving at a higher speed causes greater dynamic effects on the surface and subtrack. This results in accelerated wear, to which rails, fastenings and bottom rails are particularly exposed. In rails, cracks, wavy wear, and contact-fatigue damage are more common, in the attachments there is a lower clamping force, and in the ballast there are also crushed cavities and its crushing. The speed of travel should be predicted at the planning and design stage, so as to counteract the abovementioned phenomena already at these stages [23].

2.2. Methods

In the course of analysis of the dynamic impact of a rail vehicle on the railway surface, an original computational model was created, allowing determination of the impact of differentiated rail support on the dynamic response of the entire structure. The results of numerical calculations were verified in field tests carried out on the actual railway line in places where there was a change in the stiffness of the elements of the railway line, loaded with vehicles of different weights, moving at different speeds. The adopted research method is a theoretical model and uses experimental verification.

2.2.1. Theoretical Model

As a starting point for development of a model of a dynamically loaded railway surface, the Bernoulli-Euler beam, located on the elastic Winkler substrate, was adopted. Since the railway track has a symmetrical structure, half of its construction was considered, i.e., one rail track and the load falling on it, as well as the load-bearing system. All the supporting layers of the structure below the rail were parameterized by substitute coefficients of elasticity and damping. The model takes into account the vertical displacement of the rail caused by a multiaxis rail vehicle with a constant axle load moving uniformly. A simplification of the mapping of a rail vehicle to a moving concentrated force with constant pressure was adopted. The validity of this approach has been proven, inter alia, in [24] in which a number of variants of rail vehicle mapping with varying levels of complexity were analysed and it was found that the method of load mapping does not have a significant impact on the obtained displacement values of the railway engineering object. The same representations of a moving rail vehicle were also used in [25,26], each time giving correct, real results.

The model in this paper also takes into account the fact that in individual cross-sections, the ground layer and the railway surface can be characterized by variable values, resulting from the use of different materials and different design solutions [10,27]. The model under consideration is presented in Figure 2.

After carrying out the appropriate analysis and after taking into account the data depending on position and time, the differential equation of the deflection line of the rail course (3) was obtained. This is a differential equation of the fourth order for the spatial variable and the second order for time.

$$\frac{\partial^2 w(x,t)}{\partial t^2} \cdot m_s + \frac{\partial w(x,t)}{\partial t} \cdot c(x) = -\frac{\partial^4 w(x,t)}{\partial x^4} \cdot (E \cdot J)(x) - w(x,t) \cdot k(x) + \frac{P(x,t)}{dx}. \quad (3)$$

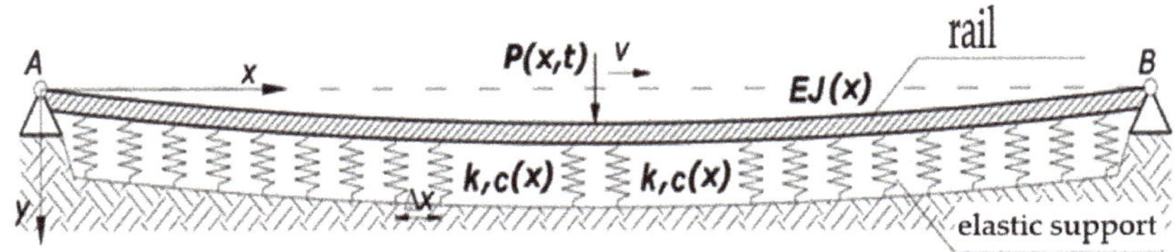

Figure 2. Rail line model.

The vertical displacement of the rail will depend on the place where the load is applied and on the time that has elapsed since the load was applied at a given point. The coefficients of the Equation (3) are described earlier.

The relation presented above was the basis for further numerical analyses. The finite differences method was used to solve it. It makes it possible to solve differential equations by replacing differential operators with difference operators. They are defined on a set of points, called a grid. Grid elements are nodes. The considered rail element was divided into sections of a length "Δx" and then, the values of individual quantities in the grid nodes were calculated. With a sufficiently high density of such a division, it is possible to achieve very accurate results, converging to those provided by the analytical solution. Thanks to this assumption, the initial extended differential equation is replaced by a system of algebraic equations [28]. To derive formulas for difference operators, the assumption was used that the desired values of the function between individual nodes are connected by fragments of the parabola. Difference schemes occurring in the finite differences method refer to the time–space grid with the parameters "Δt" and "Δx". The solutions in nodes, defined by coordinates (x, t), are calculated.

In order to solve the differential Equation (3) by the finite differences method, taking into account dynamic loads, it was necessary to determine the boundary conditions and the initial conditions of the issue under consideration. The following assumptions have been introduced:

- the rail track is freely supported at each of the ends;
- the rail track has been mentally divided into "n" sections of a length "Δx";
- the number of nodes located on the rail course is equal to "n − 1";
- from the outside of each of the support nodes, one fictitious node and one fictitious "Δx" section were added;
- the total number of nodes into which the model has been divided, including support nodes (two pieces) and fictitious nodes (two pieces), is "n + 3".

In addition, to start the numerical calculation process, one artificial time layer was introduced, and for each subsequent time layer of the calculation process, two artificial nodes for the right and left boundary conditions were introduced. The correctness of such an assumption was confirmed in the calculation process. The time-space grid with computational nodes along with the initial and boundary conditions is shown in Figure 3.

The following boundary conditions were assumed: (1) the vertical displacement in the first "fictitious" node marked by the number "1" is equal to the displacement in the first node on the rail course marked by the number "3" and taken with a minus sign, (2) the vertical displacement in the first support node marked by the number "2" is equal to 0, 3), the vertical displacement in the last support node marked by the number "n + 2" is equal to 0, 4), and the vertical displacement in the last "fictitious" node marked by the number "n + 3" is equal to the displacement in the last node on the rail course marked by the number "n + 1" and taken with a minus sign.

To determine the initial conditions, the following assumptions were made: (1) the value of the vertical displacement of a given node at the initial moment is equal to 0, (2) at the initial moment, the first node on the rail course marked by the number "3" begins to vibrate at a certain low speed and low acceleration, the values of which depend on the value of the load.

node number	1	2	3	4	5	...	i-2	i-1	i	i+1	i+2	...	n	n+1	n+2	n+3	
time	fictitious node	support node	nodes on rail												support node	fictitious node	
-1	✕	0	w_p	0	0	...	0	0	0	0	0	...	0	0	0	✕	initial
0	0	0	0	0	0	...	0	0	0	0	0	...	0	0	0	0	conditions
1	$-w_3^1$	0	w_3^1	w_4^1	w_5^1	...	w_{i-2}^1	w_{i-1}^1	w_i^1	w_{i+1}^1	w_{i+2}^1	...	w_n^1	w_{n+1}^1	0	$-w_{n+1}^1$	
...	...	0	...	...	...	...	...	...	...	...	...	...	...	...	0	...	
j-1	$-w_3^{j-1}$	0	w_3^{j-1}	w_4^{j-1}	w_5^{j-1}	...	w_{i-2}^{j-1}	w_{i-1}^{j-1}	w_i^{j-1}	w_{i+1}^{j-1}	w_{i+2}^{j-1}	...	w_n^{j-1}	w_{n+1}^{j-1}	0	$-w_{n+1}^{j-1}$	
j	$-w_3^j$	0	w_3^j	w_4^j	w_5^j	...	w_{i-2}^j	w_{i-1}^j	w_i^j	w_{i+1}^j	w_{i+2}^j	...	w_n^j	w_{n+1}^j	0	$-w_{n+1}^j$	
j+1	$-w_3^{j+1}$	0	w_3^{j+1}	w_4^{j+1}	w_5^{j+1}	...	w_{i-2}^{j+1}	w_{i-1}^{j+1}	w_i^{j+1}	w_{i+1}^{j+1}	w_{i+2}^{j+1}	...	w_n^{j+1}	w_{n+1}^{j+1}	0	$-w_{n+1}^{j+1}$	
...	...	0	...	...	...	...	...	...	...	...	...	...	...	...	0	...	
m	$-w_3^m$	0	w_3^m	w_4^m	w_5^m	...	w_{i-2}^m	w_{i-1}^m	w_i^m	w_{i+1}^m	w_{i+2}^m	...	w_n^m	w_{n+1}^m	0	$-w_{n+1}^m$	
	boundary condition I	boundary condition II	vertical displacement in particular node in particular time												boundary condition III	boundary condition IV	

Figure 3. Differential time–space grid—authors.

Thus, in order to start the numerical calculation procedure, it was assumed that the initial value of the vertical displacement of each of the nodes, resulting from the assumed initial conditions, is equal to the initial value at the time of time two steps earlier than the moment when the load is applied. This value depends on: (1) load value, (2) geometric and material characteristics of the rail, (3) surface damping value, and (4) adopted values of the time–space grid of the finite differences method, i.e., (a) spatial step and (b) temporal step.

It is worth emphasizing here that the assumed numerical value of the initial condition is used only to start the calculation process and has a negligible impact on further results and vertical displacements of the rail course calculated only using this value without rolling stock load would be negligibly small, which was confirmed in the calculations carried out.

In the calculations, there are a number of coefficients that ensure the compliance of the material parameters of individual elements of the rail surface and the position of the loading forces with the current position of the loading vehicle on the analyzed rail section. The introduced coefficients are:

- coefficient of variation of rail support (z_1);
- coefficient of effective stiffening of the rail (n_1);
- coefficient of change of a damping value (n_2);
- coefficient of occurrence of load and damping (H_i^j).

Such a set of coefficients made it possible to describe each cross-section at any time with a precisely defined set of data. These coefficients and the others that occur in the calculation process are shown in Table 2.

Particular attention should be paid to the coefficient of effective stiffening of the rail, which reflects the method of fastening the rail in various structural solutions of the railway surface. It was introduced after defining the phenomenon that the effective bending stiffness of the rail is different for ballast and ballastless surfaces [9]. In places where there is a ballast surface, its value is assumed as 1.00; in places where there is a ballastless surface, the value of 1.30 and for sheath rail systems (for example, the ERS system), where the rail is immersed in an elastic mass that limits its freedom of bending, the value of this coefficient is equal to 1.50. It should be noted that these are proprietary values, verified during the in situ research, but may require clarification and detailing with more measurements and analyses.

An analysis of the convergence of spatial and temporal steps was also performed. On its basis, the value of the spatial step Δx at the level of 0.05 m and the time step at a level lower than the critical time, which was defined as:

$$\Delta t_{kr} = \frac{2\Delta x^2}{\pi} \cdot \sqrt{\frac{m_s}{1.50 \cdot E \cdot J}} \ . \tag{4}$$

Table 2. A list of parameters that must be defined before starting the calculation.

Name of Physical Quantity	Description	Unit of Measurement
length of the rail section to be tested	L	[m]
length of surface of a given type	L_{n_i}	in the case of a homogeneous type of surface $L_{n_i} = L$
number of "Δx" sections into which the rail track has been divided	n	dimensionless (natural number must be given)
numerical duration of the analysis	T_A	[s]
substrate elastic modulus	k	[MPa]
coefficient of variation of rail support	$z_1, z_2, \ldots, z_{i,}$	dimensionless
longitudinal elastic modulus of rail steel	E	[GPa]
geometric moment of inertia of the rail cross-section	J	$[\text{cm}^4]$
coefficient of effective stiffening of the rail	n_1	dimensionless
rail mass related to the length unit	m_s	[kg/m]
level-damping coefficient	c	$[\text{Ns/m}^2]$
coefficient of change of a damping value	n_2	dimensionless
speed of passage of the rail vehicle	v	[km/h]
axle load of the rail vehicle	P	[kN]
wheelbase of the rail vehicle	R	[m]

As a result of the performed analyses and appropriate transformations, a relation was obtained, which determines the value of the vertical displacement of a given node "i" in the next considered time moment "j + 1":

$$w_i^{j+1} = \frac{1}{S_1} \cdot \{ \Delta t^2 \cdot (S_4 \cdot w_{i-2}{}^j + (-4 \cdot S_4) \cdot w_{i-1}{}^j + (6 \cdot S_4 - z_i \cdot k_i) \cdot w_i{}^j + (-4 \cdot S_4) \cdot w_{i+1}{}^j$$
$$+ S_4 \cdot w_{i+2}{}^j + H_i^j \cdot P_1) - S_2 \cdot w_i^{j-1} - S_3 \cdot w_i^j \} \tag{5}$$

$$S_1 = m_s + \frac{3}{2} \cdot n_2 \cdot H_i^j \cdot c_i \cdot \Delta t \tag{6}$$

$$S_2 = m_s + \frac{1}{2} \cdot n_2 \cdot H_i^j \cdot c_i \cdot \Delta t \tag{7}$$

$$S_3 = -2 \cdot (m_s + n_2 \cdot H_i^j \cdot c_i \cdot \Delta t) \tag{8}$$

$$S_4 = -\frac{n_1 \cdot E \cdot J}{\Delta x^4} \tag{9}$$

$$P_1 = \frac{P}{\Delta x} \tag{10}$$

The scheme of operation of the created algorithm in relation to individual nodes "i" and time moments "j" is presented in the differential time—space grid shown in Figure 3. An explicit method of solving differential equations was used [29], in which it is important to properly give the initial conditions of the calculation process.

An explicit method of searching for solutions on the time–space grid of the finite differences method was used [29]. The new value w_i^{j+1} (orange in Figure 3) is based on the values already known, calculated in the previous steps (green). The first unknown to be calculated is $w_3{}^1$. This is the vertical displacement of the first node located on the rail course at the first considered time moment (orange color). For the first calculation, therefore, five values from the time moment (j = 0, green color) and one fictitious value from the time moment (j = −1, green color) will be used. This shows how important it is to correctly give the initial condition, because otherwise any data taken into account for the first calculation would be equal to 0.

The horizontal dimension of the grid is equal to "n + 3". The vertical dimension of the grid is marked as the value "m" and corresponds to the number of time steps by which the analysis of the vertical displacement of the rail line in individual nodes will be carried out. It is important that the numerical duration of the analysis allows the value of the maximum displacement of each point to which the dynamic load is applied to be determined. It is also important to determine the number of time steps "Δt" of the whole calculation process. This quotient should be rounded up to the whole to obtain an integer and also increased by 2 to take into account the time moments marked in Figure 3 as "-1" and "0".

Figure 4 shows a flowchart of the described solution algorithm. In an orderly way, it illustrates the subsequent steps necessary to obtain the final result, which are the values of vertical displacement of the rail in individual nodes "i" of the model and in individual time moments "j", caused by the dynamic impact of the rail vehicle on the railway surface.

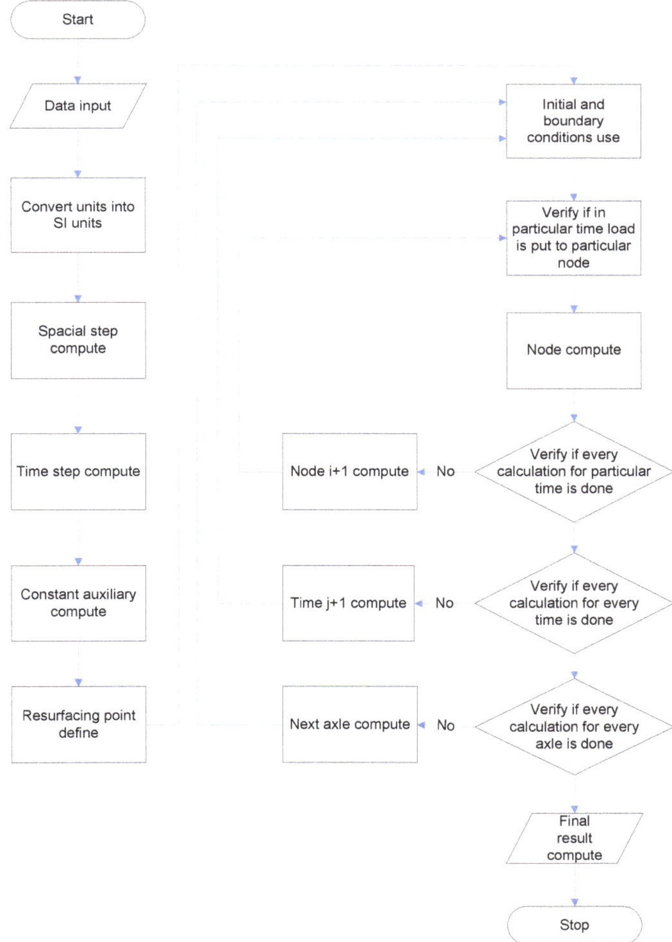

Figure 4. Block diagram of the algorithm.

2.2.2. Experimental Verification

The theoretical model should reliably describe reality so that in similar cases it is not necessary to perform field research, but only to use the theoretical modeling. In order to verify and validate the created algorithm, in situ studies of vertical displacements of dynamically loaded railway rail were carried out. The research was carried out using the laser scanning technology. The advantages of measurements performed in this technology include: high accuracy and automation, as well as the speed of measurements and the lack of the need to destroy the tested object or exclude it from operation. Among the disadvantages are: large volume of data and their long processing time, and the inability to conduct measurements during bad weather conditions, as well as the high price of the scanner and its sensitivity to mechanical damage [30]. The choice of research technology was preceded by an analysis of the literature, which gave grounds for obtaining satisfactory accuracy of the results.

Studies of vertical rail displacement caused by a passing train using the laser scanning technology were carried out in cooperation with PKP (PKP Polskie Linie Kolejowe S.A. Railway Lines Plant in Ostrów Wielkopolski, Poland) at the Railway Lines Plant in Ostrów Wielkopolski. The measurements were carried out in tracks located within two engineering structures and their access zones. In order to analyze the impact of the threshold effect on the dynamic impact of the rail vehicle on the railway surface, the following were selected: (1) place with a uniform type of surface on the object and on the trail and (2) place where there is a jumping change in the type of surface in front of and behind the object.

The research was possible thanks to cooperation with P.P.H.WObit E.K.J.Ober s.c. (https://wobit.com.pl/en) (accessed on 11 March 2022). This company made the laser scanner of the scanCONTROL LLT2610-50 series available for testing free of charge under a loan agreement concluded with the Faculty of Civil Engineering and Geodesy of the Military University of Technology in Warsaw. The measuring set included the following elements: (1) laser scanner of the scanCONTROL LLT2610-50 series, (2) a tripod with mounting screws enabling permanent fixation of the measuring device, (3) a power set, and (4) a cable enabling the scanner to be connected to a computer. The scanner allows to measure up to 2,560,000 points per second at a frequency of up to 4000 Hz. Measurement accuracy is 2 μm. The scanner is enclosed in a metal casing, which ensures an IP65 degree of protection. At the bottom of the scanner there is a light source and a receiver. There are two connectors on the housing: (1) Ethernet (used to communicate with a PC or Plc controller) and (2) multifunctional (used, among others, for power supply). To communicate the device with a computer, a cable with an RJ45 connector on one side and a connector dedicated to the scanner on the other [31] is used. For the proper operation of the entire measuring system, one needs a computer with the appropriate software installed and a power generator providing a constant power source. Before starting measurements, it is necessary to check the correctness of the scanner's connection with the computer, select the appropriate software configuration and assign it to the device, as well as define the scanner's work settings by setting parameters such as exposure time and the number of measured profiles per second. In selection of the correct parameters, a preview of the measurement, which is displayed in real time, in one of the software modules, is helpful. In addition, one must specify a name and location to save the result file. The recording is made automatically after the measurements are completed. The scanner measuring set is shown in Figure 5.

The scanner was placed transversely to the track axis, so that the infrared radiation it sent was directed at the rail foot. The holder with the scanner was each time mounted on a metal rod about 0.65 m long driven into the railway crushed stone in a way that prevented any movement of the measuring system. The measuring stations are shown in Figures 6–9.

Figure 5. The scanner measuring set.

Figure 6. Measuring station—railway viaduct—km 83,808 LK No. 272.

Figure 7. Measuring station—jumping change in the stiffness of the surface—access to the railway viaduct at km 88,882 LK No. 272.

Figure 8. Railway viaduct at km 83,808 LK No. 272—measurement of the vertical displacement of a rail dynamically loaded with a train of Koleje Wielkopolskie.

Figure 9. Railway viaduct at km 88,882 LK No. 272—measurement of the vertical displacement of a rail dynamically loaded with a freight train.

3. Results and Discussion

3.1. General Comparisons of Calculation Results to In Situ Test Results

The results of the measurements were compared to the results of calculations obtained from the developed theoretical model for given surfaces and vehicles that occurred during the field tests. The results are summarized in Table 3.

Differences of up to 3.5% were obtained between the results. A satisfactory convergence of the results obtained in in situ measurements with the results calculated using the developed algorithm and the finite differences method was observed. The characters of the charts are in all the cases the same. It was noticed that the curve mapping the measured vertical displacements of the rail line is more irregular, while the calculation curve is smoothed and much more symmetrical. In the case of calculations for low speeds of the order of 30 km/h (measurement point 4), a higher density of results was noted due to the need to adopt a longer analysis time and a denser time step. In the case of measurements for high speeds, of the order of more than 100 km/h (measuring point 5) an overlap of displacements caused by adjacent axes was noted. This phenomenon is caused by the relatively short wheelbase in the bogie of the rail vehicle and the short time passing between successive loads. This issue, however, did not affect the recorded maximum value of the vertical displacement of the rail.

Table 3. Comparison of results obtained in in situ measurements to the theoretical results received from the developed algorithm using the finite differences method.

The Maximum Value of the Vertical Displacement of the Rail Due to the Dynamic Load				
Measuring Point Number	**Measuring Point**	**Measurement Value [mm]**	**Calculation Value [mm]**	**Difference**
1	uniform ballast surface	1.256	1.232	−1.9%
2	uniform ballast surface	0.968	1.002	+3.5%
3	uniform ballast surface	0.998	0.982	−1.6%
4	ballast surface in front of the object and ballastless on the object	1.393	1.398	+0.4%
5	ballast surface in front of the measuring point (wooden sleepers) and ballast surface (concrete sleepers) behind the measuring point	1.317	1.302	−1.1%

At the same time, factors that may have a direct impact on the results of measurements and calculations must be indicated. The accuracy of the measurements carried out is decisively influenced by the adoption of the appropriate parameters of the laser scanner, such as: length and frequency of "exposure". It is also important to fix the device in such a way that it remains in a constant position throughout the measurements.

In the context of the accuracy of calculations, the parameters of the computer used for the analysis are of great importance. The better they are, the more it is possible to assume a denser division of the model into nodes and time moments, which directly translates into the accuracy of calculations. In addition, it is important to adopt appropriate numerical values of data and input parameters for the railway surface. While in the case of such attributes as the mass of the rail, the modulus of longitudinal elasticity of rail steel or the geometric moment of inertia of the rail cross-section, choosing the right values does not raise doubts, in the case of the modulus of elasticity of the substrate or the damping coefficient, it is much more difficult to assume the right quantities. In this work, theoretical values determined based on the analysis of the literature on the subject were used, which, however, was confirmed by the high convergence of the results of calculations and measurements.

The influence of the dynamic impact of a rail vehicle on the railway surface within the transition zones in front of and after the engineering object, depending on its design, was analyzed.

As a starting point, a combination of a ballast and ballastless surface was adopted, where there is a step change in the parameters of the structure. A ballast surface with concrete sleepers and a good subtrack surface, as well as a ballastless surface in the rail system in the ERS type sheath, were adopted. Based on the obtained results, it is clear that within the transition zone, the dynamic effects on the surface are much greater and weaken rapidly when changing the type of structure.

For comparison with the above results, a solution was analyzed in which the transition zone in front of the engineering object was strengthened, which was reflected in the form of a 50% higher value of the elastic modulus of the substrate. Based on the obtained results, smaller differences in dynamic effects on the surface were noticed at the tested points.

The effect of the gradual change in the elasticity of the rail support within the transition zone on the magnitude of the dynamic impact of the rail vehicle on the surface was also analyzed. It was assumed that the modulus of elasticity of the support increases evenly every one meter over a length of $L_{n1} = 10$ m in front of the object from the value characteristic for the ballast surface equal to $k_{1.0} = 45$ MPa to the value characteristic for the ballastless surface equal to $k_{2.0} = 99$ MPa. In this case, a gradual and mild decrease in the magnitude of dynamic effects on the surface in subsequent fragments of the model was noticed.

In order to compare the impact of the reinforcement of transition zones and the gradual change in the elasticity of the rail support on the reduction of the magnitude of dynamic interactions on the railway surface, the calculated results were tabularly compiled. Table 4 contains the maximum values of vertical displacement of the rail at critical points within the place of change of the type of surface (from ballast to ballastless) i.e., (1) just before the point of change and (2) just after the point of change.

Table 4. Comparison of three variants of the construction of the railway surface within the zone in front of the engineering facility.

The Maximum Value of the Rail Vertical Displacement Due to the Dynamic Load Depending on the Structure of the Railway Surface within the Zone in Front of the Engineering Object			
Location (Relative to the Place of Change of the Type of Surface from Ballast to Ballastless)	**Step-by-Step Change of Structure Parameters [mm]**	**Reinforcement of Transition Zones [mm]**	**Gradual Change in the Elasticity of the Rail Support [mm]**
before	1.201	0.925	0.804
behind	0.638	0.622	0.630
difference	0.563	0.303	0.174
difference in [%]	47	33	22

By strengthening the transition zones, a 14% reduction in the magnitude of the dynamic impact on the railway surface within the site of the change of the type of surface in front of the engineering site was achieved. By applying a gradual change in the elasticity of the rail support, a 25% reduction in the dynamic effects on the railway surface within the site of the change of the type of surface in front of the engineering object was achieved.

From the analysis, it was concluded that a gradual change in the elasticity of the rail support within the transition zones of the engineering object reduces the negative impact of the dynamic impact of the rail vehicle on the railway surface and this solution is better compared to a step change in the elasticity of the support.

3.2. Detailed Analysis

Detailed results of numerical analysis are given in Figures 10–14.

3.3. Discussion

The paper analyzed the dynamic impact of a rail vehicle on various solutions of the railway surface structure, with particular emphasis on the phenomenon of the threshold effect, which occurs within the transition zones of the engineering facility. The problem of locally variable stiffness of the railway surface has been identified, which in turn may lead to accelerated degradation of the structure and the need to incur increased expenditures on maintaining the infrastructure in proper condition.

Based on the analytical and numerical considerations, a computational model was created, thanks to which it is possible to determine the impact of various variants of the rail support on the dynamic response of the entire structure. The starting point for the deliberations was the Bernoulli–Euler beam. The dynamic load caused by the passage of a multiaxle rail vehicle and different vibration-damping properties of different types of surfaces are taken into account. As a consequence, the fourth-order differential equation was obtained. It was solved by the finite differences method. A script in MATLAB was developed for a numerical solution of the problem.

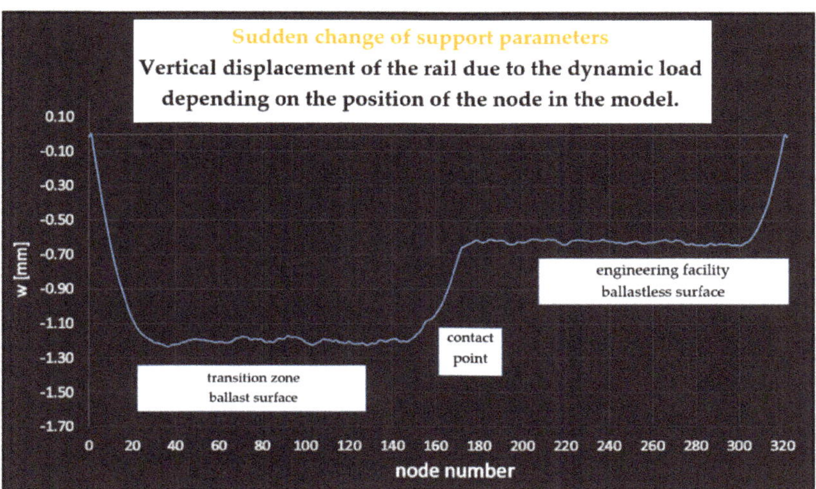

Figure 10. The dependence of the maximum vertical displacement of the rail caused by the dynamic load, depending on the position of the node in the model in the case of a step change in the parameters of the structure.

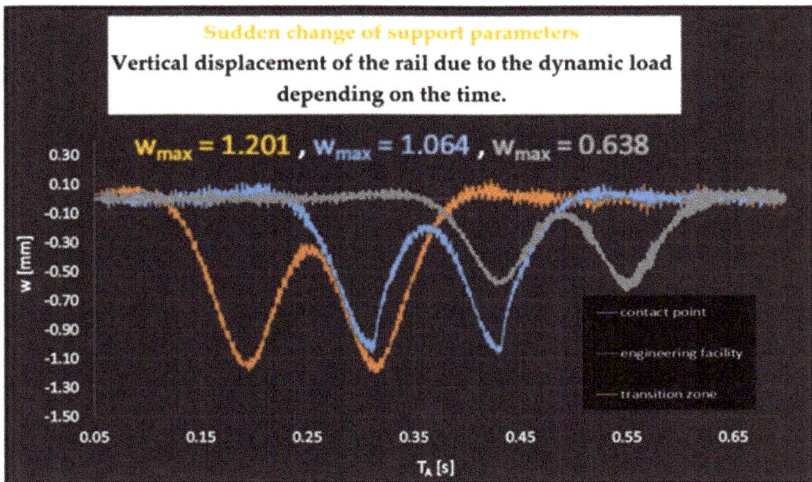

Figure 11. Vertical displacement caused by dynamic load, depending on time in the case of a step change in the parameters of the structure for three different nodes: 1. node on the transition zone (ballast surface), 2. node in the place of a jumping change in the parameters of the structure (connection of the surface), and 3. node on the engineering object (ballastless surface).

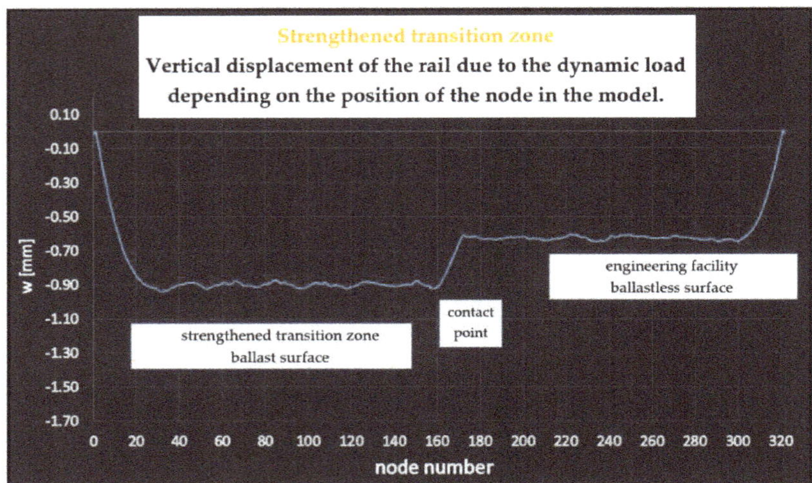

Figure 12. Dependence of the maximum vertical displacement of the rail caused by dynamic load depending on the position of the node in the model in the case of reinforcement of transition zones.

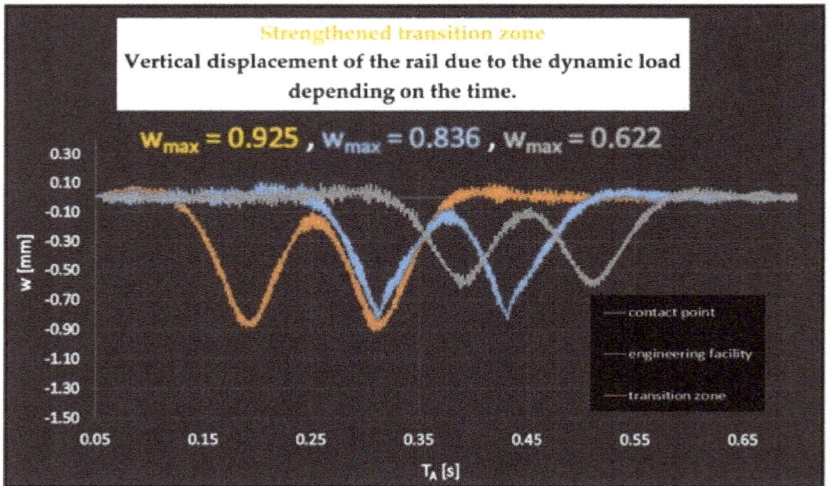

Figure 13. Vertical displacement caused by dynamic load, depending on time in the case of reinforcement of transition zones for three different nodes: (1) node on the transition zone (ballast surface), (2) node in the place of change of structural parameters (surface connection), and (3) node on the engineering object (ballastless surface).

In order to verify and validate the created algorithm, in situ studies of vertical displacements of dynamically loaded railway rail were carried out. For this purpose, laser scanning technology was used.

Thus:

(1) The effectiveness of the finite differences method in the context of solving multiparameter differential equations describing the dynamics of the surface loaded by a passing railway vehicle has been confirmed. The algorithm used allowed for precise determination of the impact of technical and operational parameters on the magnitude

of dynamic interactions of a rail vehicle on the railway surface. At the same time, it should be emphasized that development of an algorithm using the finite differences method does not involve the need to use complicated and expensive computer software.

(2)   The results obtained using the laser scanning technology are characterized by high accuracy. The most important in the correct conduct of measurements was identified as a precise fixation of the measuring device, in such a way that it remained in a constant position throughout the measurements, as well as the adoption of appropriate parameters of the laser scanner's operation, such as: the length of exposure and the frequency of measurements. The issue of the possibility of taking measurements for high speeds of trains may be a cause for concern—the maximum speed of the rail vehicle at which the measurements were made in this work was 120 km/h. However, in the light of the results obtained, the usefulness of the method used in the context of measurements of the rail surface displacements was positively assessed. Certainly, this technology can also be used in other areas where high precision and accuracy are required. At the same time, it should be noted that in the course of measurements of displacements of the railway surface caused by dynamic load, measurements of the condition of individual elements of the surface were additionally made. This issue was not the subject of analysis of this work, but it should be pointed out that the laser scanning is also useful in the assessment, detection, and identification of surface defects of rails, sleepers, and fasteners.

(3)   From the analysis, it was concluded that a gradual change in the elasticity of the rail support within the transition zones of the engineering object reduces the negative impact of the dynamic impact of the rail vehicle on the railway surface and this solution is better compared to the step change in the elasticity of the support.

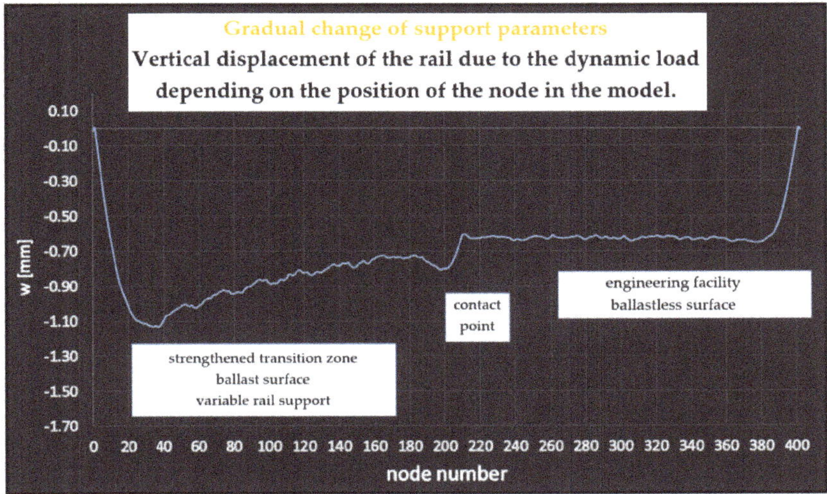

Figure 14. The dependence of the maximum vertical displacement of the rail caused by the dynamic load depending on the position of the node in the model in the case of a gradual change in the elasticity of the rail support within the transition zone.

4. Conclusions

The article combines theoretical simulations of the dynamic process of a rail response to the wheel load by a passing rail vehicle with precise tests on a real railway line, in which laser scanning with a sensor recording fast-changing processes was used.

By strengthening the transition zones, the negative effects of dynamic loads on the railway surface within the site of the change of the type of surface in front of the engineering site have been reduced by 14%.

By applying a gradual change in the elasticity of the rail support, the negative effects of dynamic loads on the railway surface within the place of change of the type of surface in front of the engineering object were reduced by 25%. Studies of concrete mixtures with different additives have already been undertaken which, in the laboratory scale, were used to make models of ballastless railway surface elements. The first results of the research indicate their possible applicability in practice.

Increasing the elasticity of the rail support within the transition zones before and after the object can be achieved by reducing the spacing of sleepers, better compaction of the ballast layer or strengthening of the upper layers of the subtrack, as well as by covering the track with one or more layers of transition plates [16,32].

It should be noted, however, that reducing the magnitude of the dynamic impact of a rail vehicle on the surface does not completely eliminate the phenomenon of the threshold effect, but only eliminates its negative impact on durability of the structure in zones where the type of surface changes [9].

The effects of the threshold phenomenon, in addition to vertical deformations of the rail, may also be other phenomena accompanying this effect, which may lead to an increase in track twisting and uneven wear of rails and damage to fastenings on both types of surfaces. Gaps may form under the sleepers, which threatens the stability of the structure. The threshold phenomenon has a negative impact not only on the railway surface, but also on the object that is exposed to excessive loads and vibrations.

Taking into account the above conclusions, areas exposed to the threshold phenomenon should be subjected to special supervision both in terms of current diagnostic activities and planned maintenance works.

In further research, in addition to the magnitude of displacements caused by the dynamic impact of a rail vehicle on the railway surface, it would also be necessary to determine the impact of these interactions on durability of the railway surface elements, such as sleeper rails and fasteners. It would certainly be valuable to determine the fatigue life of these elements in the zones of the threshold phenomenon and compare it to the nominal fatigue life for built-up elements outside the boundaries of transition zones. In addition, it would be worth identifying the processes taking place in the railway subtrack, i.e., the effects of the threshold phenomenon previously defined as "deep". The implementation of the above works would allow for a more complete knowledge of the impact of the threshold phenomenon on the condition of the railway surface and could be helpful in more effective planning of maintenance works within the transition zones in front of and behind engineering facilities.

Moreover, the load on a rail by a passing vehicle that was used in the article was movable, but constant as to the value adequately reflecting the weight of the vehicle. The comparative results are satisfactory. However, in future studies, the loading force will take into account the suspension structure of the vehicle, so its value will be determined by this suspension. Such a model was presented by authors Vahid Bokaeian, Mohammad Ali Rezvani, and Robert Arcos in [33].

Author Contributions: Conceptualization, T.L. and W.I.; methodology, T.L. and W.I.; software, T.L.; validation, W.I. and J.T.; formal analysis, T.L.; investigation, T.L.; writing—original draft preparation, T.L.; writing—review and editing, W.I. and T.R.; visualization, T.L. and D.P.; supervision, W.I. All authors have read and agreed to the published version of the manuscript.

Funding: This research was funded by Faculty of Civil Engineering and Geodesy of the Military University of Technology, Warsaw, Poland—grant UGB number 794.

Institutional Review Board Statement: Not applicable.

Informed Consent Statement: Not applicable.

Data Availability Statement: The data presented in this study are available upon request from the corresponding author.

Acknowledgments: The authors would like to acknowledge: (1) the Authority of the Faculty of Civil Engineering and Geodezy of the Military University of Technology for providing administrative support during conducting our scientific work on this article, (2) the National Infrastructure Manager in Poland—PKP Polskie Linie Kolejowe S.A. Railway Lines Plant in Ostrów Wielkopolski (Wolności 30, 63-400 Ostrów Wielkopolski) for enabling measurements to be carried out on active railway track, and (3) P.P.H. WObit E.K.J Ober s.c. (Dęborzyce 16, 62-045 Pniewy) company for providing a scanner for measurements.

Conflicts of Interest: The authors declare that they have no conflict of interest.

References

1. Basiewicz, T.; Towpik, K.; Gołaszewski, A. Nawierzchnia kolejowa z kompozytem tłuczniowym. Prace naukowe Politechniki Warszawskiej. *Transport* **2010**, *72*, 77–86.
2. Bałuch, H. *Budownictwo Komunikacyjne*; Military University of Technology: Warszawa, Poland, 2001.
3. Grulkowski, S.; Kędra, Z.; Koc, W.; Nowakowski, M. *Drogi Szynowe*; Wydawnictwo Politechniki Gdańskiej: Gdańsk, Poland, 2013.
4. Rail.one. GmbH Rail.one GmbH company Materials, Rheda 2000Ballastless Track System. Available online: https://www.railone.com/fileadmin/daten/05-presse-medien/downloads/broschueren/en/Rheda2000_EN_2011_ebook.pdf (accessed on 11 March 2022).
5. Tines sp. z o o. Materials of Tines sp. z o o. Available online: https://TINESLC-L (accessed on 14 March 2022).
6. Sybilski, D. Nawierzchnia Kolejowa z warstwami asfaltowymi. *Problemy Kolejnictwa Zeszyt* **2012**, *156*, 68–78.
7. RynekInfrastruktury.pl. Available online: http://www.rynekinfrastruktury.pl/koleje.html (accessed on 14 March 2022).
8. Sołkowski, J.; Kudła, D. Analiza Niejednorodności Mechanicznych Nawierzchni i Podtorza w Obrębie Obiektu mostowego. Zeszyty Naukowo-Techniczne Stowarzyszenia Inżynierów i Techników Komunikacji Rzeczypospolitej Polskiej, Krakó. 2016. Available online: http://yadda.icm.edu.pl/baztech/element/bwmeta1.element.baztech-96c9ea1a-cdac-4031-85d5-243da1eb1405 (accessed on 14 March 2022).
9. Sołkowski, J. Zarys analizy efektu progowego przy łączeniu nawierzchni podsypkowych z innymi typami nawierzchni. *Technika Transportu Szynowego 10/12*, 59–65.
10. Sołkowski, J. *Efekt Progowy w Nawierzchniach Szynowych*; Politechnika Krakowska: Warszawa, Polska, 2013.
11. Halliday, D.; Resnick, R.; Walker, J. Podstawy Fizyki, Tom 2. Wydawnictwo Naukowe PWN, Warszawa. 2008. Available online: https://ksiegarnia.pwn.pl/Podstawy-fizyki-Tom-2,68432815,p.html?gclid=EAIaIQobChMIi5iimJrF9gIVxgh7Ch30BAS8EAQYAiABEgLx6PD_BwE (accessed on 14 March 2022).
12. Vostroukhov, A.; Metrikine, A. Periodically supported beam on a visco-elastic layer as a model for dynamic analysis of a high-speed railway track. *Int. J. Solid Struct.* **2003**, *2003*, 5723–5752. [CrossRef]
13. Kerr, A. The Determination of the Track Modulus k for the Standard Track Analysis. Department of Civil and Environmental Engineering University of Delaware: Newark, NJ, USA, 2002.
14. Powrie, W.; Le Pen, L. *A Guide to Track Stiffness*; Cross Industry Track Stiffness Working Group, University of Southampton: Southampton, UK, 2016; ISBN 9780854329946.
15. Selig, E.; Dingqing, L. Track modulus: Its meaning and factors influencing it. *Trans. Res. Record* **1994**, *1470*, 47–54.
16. Skrzyński, E. Podtorze kolejowe na liniach dużych prędkości. *Problemy Kolejnictwa Zeszyt* **2013**, *161*, 87–125.
17. Yang, Y.; Yau, J.; Yao, Z.; Wu, Y. Vehicle-bridge interaction dynamics: With applications to high-speed railways. *Civil Eng. Knowl. Base* 2004.
18. Lewandrowski, T.; Idczak, W.; Muzolf, P. *Przybliżone Modelowanie Układu "Pojazd Szynowy-Nawierzchnia-Podłoże"*; Military University of Technology: Warszawa, Poland, 2017.
19. Uzarski, D. *Railroad Track Design*; University of Illinois: Chicago, IL, USA, 2009.
20. Starczewski, Z. *Drgania Mechaniczne*; Politechnika Warszawska: Warszawa, Polska, 2010.
21. PKP Polskie Linie Kolejowe, S.A. Wykaz maksymalnych nacisków osi. Warszawa. 2015. Available online: https://www.plk-sa.pl/files/public/user_upload/pdf/Reg_przydzielania_tras/Regulamin_2016_2017/09.10.2017/N_ZAL_2.2_20171005134828.pdf (accessed on 14 March 2022).
22. Raczyński, J. Nowy rekord świata pociągu TGV-578,8 km/h. *Technika Transportu Szynowego* **2007**, *5–6*, 15–35.
23. Towpik, K. Utrzymanie nawierzchni na liniach dużych prędkości jako element ryzyka w procesie eksploatacji. *Technika Transportu Szynowego* **2013**, *20*, 77–80.
24. Szafrański, M. *Wpływ Sposobu Odwzorowania Pojazdu Szynowego na Odpowiedź Dynamiczną Przęsła Mostowego*; Infrastruktura Transportu Szynowego: Gdańsk, Poland, 2019.
25. Kaewunruen, S.; Lewandrowski, T.; Chamniprasart, K. Nonlinear modeling and analysis of moving train loads on interspersed railway tracks. In Proceedings of the 6th ECCOMAS Thematic Conference on Computational Methods in Structural Dynamics and Earthquake Engineering, Rhodes Island, Greece, 15–17 June 2017.
26. Kaewunruen, S.; Lewandrowski, T.; Chamniprasart, K. Dynamic responses of interspersed railway tracks to moving train loads. *Int. J. Struct. Stab. Dyn.* **2018**, *18*, 1850011. [CrossRef]

27. Ataman, M. *Analiza Drgań Nawierzchni i Podtorza pod Wpływem Obciążeń Ruchomych z Dużymi Prędkościami*; Oficyna Wydawnicza Politechniki Warszawskiej: Warszawa, Polska, 2019.
28. Cichoń, C.; Cecot, W.; Krok, J.; Pluciński, P. Metody Komputerowe w Liniowej Mechanice Konstrukcji. *Politechnika Krakowska.* 2009. Available online: https://www.tu.kielce.pl/~{}sk/files-epi/epi-metody-komputerowe.pdf (accessed on 14 March 2022).
29. Kincaid, D.; Cheney, W. Numerical Analysis. In *Mathematics of Scientific Computing*; University of Texas: Austin, TX, USA, 2002.
30. Wasiuk, R.; Szadkowski, A.; Mahrburg, A.; Szadkowska, Ż. Mobilne skanowanie laserowe obiektów liniowych. *Drogownictwo* **2011**, *11*, 360–365.
31. WObit. *WObit Company Materials*; Wobit: Pniewy, Poland, 2019.
32. Surowiecki, A.; Duchaczek, A.; Saska, P. *Bezpieczeństwo Techniczne Toru Kolejowego w Szczególnych Warunkach Eksploatacji*; Czasopismo Logistyka: Wrocław, Poland, 2015.
33. Bokaeian, V.; Rezvani, M.A.; Arcos, R. Nonlinear impact of traction rod on the dynamics of a high-speed rail vehicle carbody. *J. Mechan. Sci. Technol.* **2020**, *34*, 4989–5003. [CrossRef]

Article

Dynamic Characteristics of a Traction Drive System in High-Speed Train Based on Electromechanical Coupling Modeling under Variable Conditions

Ka Zhang [1,2], Jianwei Yang [1,2,*], Changdong Liu [1,2], Jinhai Wang [1,2] and Dechen Yao [1,2]

[1] School of Mechanical-Electronic and Vehicle Engineering, Beijing University of Civil Engineering and Architecture, Beijing 100044, China; zhangka13230105622@163.com (K.Z.); 1108140721004@stu.bucea.edu.cn (C.L.); wangjinhai@bucea.edu.cn (J.W.); yaodechen@bucea.edu.cn (D.Y.)

[2] Urban Rail Transit Vehicle Service Performance Guarantee Key Laboratory of Beijing, Beijing University of Civil Engineering and Architecture, Beijing 100044, China

* Correspondence: yangjianwei@bucea.edu.cn

Abstract: The traction drive system of a high-speed train has a vital role in the safe and efficient operation of the train. This paper established an electromechanical coupling model of a high-speed train. The model considers the interaction of the gear pair, the equivalent connecting device of the transmission system, the equivalent circuit of the traction motor, and the direct torque control strategy. Moreover, the numerical simulation of the high-speed train model includes constant speed, traction, and braking conditions. The results indicate that the meshing frequency and the high harmonics rotation frequency constitute the stator current. Furthermore, both frequencies are evident during constant speed. However, they are blurry among other conditions except for twice the rotation frequency. Meanwhile, the rotor and stator currents' root-mean-square (RMS) values during traction are less than the RMS value during braking. The initiation of traction and braking causes a significant increase in current. During the traction and braking process, the RMS value of the current gradually decreases. Therefore, it is necessary to pay attention to the impact of the transition process on system reliability.

Keywords: dynamic characteristics; traction drive system; direct torque control; electromechanical coupling modeling; variable conditions

Citation: Zhang, K.; Yang, J.; Liu, C.; Wang, J.; Yao, D. Dynamic Characteristics of a Traction Drive System in High-Speed Train Based on Electromechanical Coupling Modeling under Variable Conditions. *Energies* **2022**, *15*, 1002. https://doi.org/10.3390/en15031202

Academic Editors: Larysa Neduzha, Jan Kalivoda and Antonio Rosato

Received: 13 December 2021
Accepted: 3 February 2022
Published: 7 February 2022

Publisher's Note: MDPI stays neutral with regard to jurisdictional claims in published maps and institutional affiliations.

Copyright: © 2022 by the authors. Licensee MDPI, Basel, Switzerland. This article is an open access article distributed under the terms and conditions of the Creative Commons Attribution (CC BY) license (https://creativecommons.org/licenses/by/4.0/).

1. Introduction

In high-speed trains, the traction transmission system transforms the electric energy supplied to the power grid into the mechanical energy of train movement by transferring torque, providing a stable power source for the train [1]. The traction drive system of a high-speed train is a complex electromechanical coupling system [2], which consists of a traction inverter, traction motor, and gear transmission system. The complex operating environment and high-intensity working state often cause unavoidable damage to the train traction drive system. For instance, Kuznetsov et al. [3] studied the influence of external factors on the dynamic performance of the traction motor internal structure; Wang et al. [4,5] studied the influence of wheel damage on the dynamics of gear pair and gearbox. Henao et al. [6] studied the influence factors of railway vehicle gearbox failure under the joint action of internal and external factors. The failure of a traditional traction system causes hidden trouble to train operation safety. Therefore, further study of dynamic characteristics of traction drive systems under complex working conditions is of great significance to the parameter optimization of dynamic performance index and operation safety of the high-speed train.

In the past few decades, many scholars have researched the dynamic characteristics of train transmission systems from mechanical modelling. Garg et al. [7] independently studied the dynamic characteristics of vehicle operation, ignoring track influence. Subsequently, Zhai and Sun et al. [8] proposed a classic vehicle–track coupling dynamic model to comprehensively consider the wheel–rail interaction's dynamic characteristics. Many scholars have carried out in-depth studies on the dynamic interaction between vehicle and track systems. Ren et al. [9] studied the lateral dynamic characteristics of the vehicle turnout system. Nelson et al. [10] studied the vertical dynamic characteristics between train and track. Then, Wang et al. [11] studied the lateral dynamic characteristics of the wheel–rail system in curved sections. These research achievements have promoted the development of railway vehicle dynamics research extensively. In order to make the research work restore the actual situation of the vehicle system, more and more scholars gradually introduce the gear transmission system into the established mechanical model. Wang and Yang et al. [12] studied the nonlinear behavior of the gear system of the railway vehicle, considering time-varying random excitation and ignoring rail irregularities. On this basis, Wang et al. [13] studied the dynamic behavior of railway vehicle gear–wheelset systems under traction/braking conditions. The numerical calculation results show an important relationship between vehicle time-frequency dynamic characteristics, slip speed, and wheel–rail nonlinear interaction force.

Huang et al. [14] established a vehicle mechanical structure model considering the gear transmission system, and the vehicle vibration was mainly studied from the perspective of the internal incentive problem. Zhang et al. [15,16] focused on wheel–track coupling conditions and saturated wheel–rail contact influence on vehicle resonance characteristics. The results show that vehicle vibration has an inseparable relationship with wheel adhesive properties. Yao et al. [17] also proved this result. Chen et al. studied different influencing factors, such as tooth profile meshing excitation [18], dynamic transmission error [19], and the tooth root cracks' propagation mechanism of cylindrical gears [20,21]. Based on the gear research above, a locomotive gearbox model was established to study the dynamic characteristics of the locomotive gear transmission effect under traction conditions [22].

Wang and Yang et al. [23,24] studied the influence profile shift on the time-varying meshing stiffness and dynamic characteristics. They also studied the dynamic characteristics of railway vehicle axlebox bearings under fault conditions, which has provided help for the design and fault diagnosis of gears and bearings. Wang et al. [25] established a vehicle–track coupling dynamic system containing gearboxes under variable speed conditions. The study found that the gearbox and vehicle–track coupling systems have apparent dynamic interactions. Furthermore, Liu et al. [26] studied the sliding dynamic performance of rolling bearings under acceleration conditions. They made a reasonable prediction of bearing slipping under load. The vehicle dynamic model's traction motor structure is often simplified as a rigid mechanical body due to the aforementioned gear transmission system. Therefore, the only role of the traction drive system is to transmit torsional vibrations.

On the other hand, the influence of traction motors on the dynamic characteristics of railway vehicles has also made significant progress. By studying the vibration and noise of the motor system, Kim et al. [27] concluded that harmonic components have an inseparable relationship with the acceleration of the mechanical system. Qi and Dai [28] established a vehicle model for high-speed EMUS to explore the influence of motor harmonic torque on wheel wear. Under traction conditions, they found that wheel wear accelerated with increased harmonic components. Youb [29] and Pustovetov K.M. [30] proposed that the operation of three-phase asynchronous motors with unbalanced voltage supply or electromagnetic torque harmonics may cause severe damage to bearings. Thus, it heavily influences the smoothness of train operations. Wang et al. [31] studied the state of vehicles with or without wheel wear of the traction motor. The results showed that powered vehicles significantly influenced wheel wear depth and contact energy. Mei et al. [32] established a motor–track space coupling dynamics model considering the traction variable speed system. The results showed that the traction motor significantly influenced the overall vehicle dynamics characteristics and the wheel–rail contact state in the low-speed range. Wu et al. [33] studied the influence of DC-link voltage pulsation of the high-speed train transmission system on mechanical structure fatigue. The numerical analysis results found that the feedback compensation algorithm can suppress the structural vibration and prolong the structure's service life. By establishing the electromechanical coupling model of a high-speed train, Zhu et al. [34] mainly studied the influence of harmonic torque of the traction motor on the vibration of vehicle and gearbox. The research results showed that the influence of the harmonic torque of the motor on the vibration of the gearbox was different with the change of speed, and the influence was more evident at high speed. Li et al. [35] have done a lot of research on the suppression of motor harmonics to reduce the fatigue damage of the car body and gearbox effectively. The results are validated by Liu et al.'s research results [36]. In addition, with the gradual increase of train speed, high-speed trains also have hidden risks due to their huge kinetic energy during operation. In order to track and detect the speed of trains at all times, Liang et al. [37] designed a train speed tracking controller by using neural network, which greatly reduced the speed tracking error. At the same time, Xu et al. [38], through the dynamic surface method, and Hou et al. [39], through the multi-particle model method, also studied the speed tracking control. According to the above research, the train electrical system's dynamic characteristics and the mechanical system's movement and vibration have an inseparable relationship. Therefore, it is necessary to consider the influence of electric systems when studying the dynamic characteristics of train traction drive systems.

In this paper, the electromechanical coupling model of a high-speed train, including the complete transmission system, is established by considering the electrical structure and mechanical structure of the traction drive system. Subsequently, the dynamic responses of the electromechanical coupling model under variable conditions were studied by co-simulation, and the dynamic characteristics of the traction motor were revealed by numerical analysis. The simulation results improve the study of high-speed train traction drive systems' dynamic characteristics and provide engineers guidance to design and optimize parameters. The paper is organized as follows. Section 2 introduces the realization of the electromechanical coupling model of a high-speed train. Section 3 analyzes the results of model co-simulation. Furthermore, Section 4 concludes the paper.

2. Electromechanical Coupling Modeling of a High-Speed Train

A high-speed train is a typical electromechanical coupling system. This section illustrated mechanical and electrical part modeling and realizes the electromechanical coupling system.

2.1. Mechanical Model

This paper chooses a motor car of the EMU in service as the research object. Figure 1 is the vehicle dynamic model, and Figure 2 is the structural topology of the model. The car body, bogie frame, wheelset, motor, and gearbox are rigid bodies in the dynamic model. The suspension force of both the primary and secondary suspensions consists of parallel springs and damping systems. In addition, the vertical damper of the primary suspension and secondary suspension and the anti-snake movement damper and lateral stops of the secondary suspension are considered nonlinear parameters. The traction motor is fixed to the bogie frame using bolts. The gearbox is connected with the bogie frame through the hanger rod. The transmission system's rolling bearing and coupling structure is simplified as suspension force with stiffness damping. The gear pair considers the transmission torque and describes the internal excitation factors such as tooth clearance and time-varying meshing stiffness of the gear transmission model. Moreover, the driven gear is fixed to the wheel axle by restraint and maintains synchronous motion.

Figure 1. Top view of the high-speed train dynamics model.

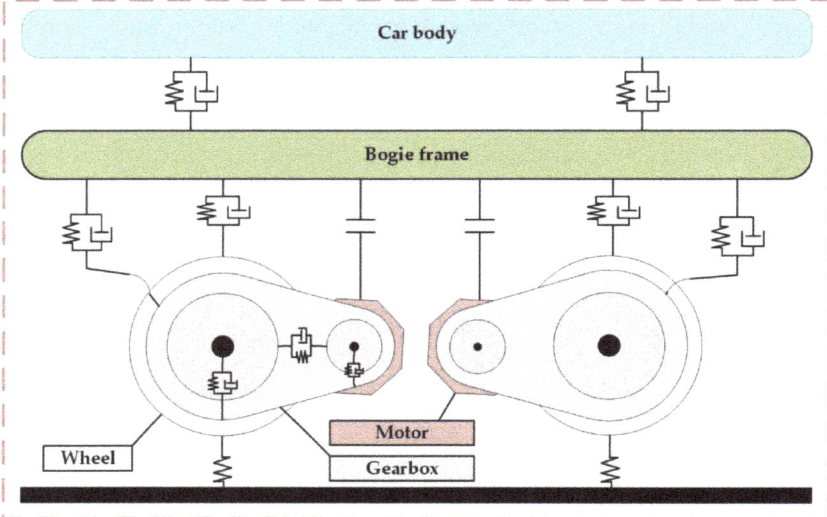

Figure 2. Topology of vehicle model structure.

In Table 1, the main parameters of cylindrical helical gear transmission in the gearbox model. Moreover, as shown in Table 2, the car body, bogie frame, and wheelset in the vehicle dynamic model have six degrees of freedom relative to the ground. The axle box, gearbox, gear, pinion, and motor rotor have a degree of freedom of rotation around the lateral direction. There are 66 degrees of freedom in the vehicle dynamic model.

Table 1. Parameters of the gear pair structure.

Parameter	Value	Parameter	Value
Gear teeth	69	Pinion teeth	29
Gear modification coefficient	0.015	Pinion modification coefficient	0.2
Modulus (mm)	7	Modulus (mm)	7
Width of the gear (mm)	70	Width of the pinion (mm)	70
Spiral angle (deg)	20	Backlash (mm)	0
Teeth stiffness ratio	0.8	Poisson's ratio	0.3
Young's modulus (Pa)	2×10^{11}	Damping coefficient (Nm/s)	5000

Table 2. Degree of freedom of each structure in a vehicle model.

Parts	Longitudinal	Lateral	Vertical	Roll	Yaw	Pitch	Note
Car body	X_c	Y_c	Z_c	Φ_c	θ_c	Ψ_c	
Bogie frame	X_{fi}	Y_{fi}	Z_{fi}	Φ_c	θ_c	Ψ_c	$i = 1\text{--}2$
Wheelset	X_{wi}	Y_{wi}	Z_{wi}	Φ_{wi}	θ_{wi}	Ψ_{wi}	$i = 1\text{--}4$
Axle box					θ_{ai}		$i = 1\text{--}8$
Gearbox					θ_{bi}		$i = 1\text{--}4$
Gear					θ_{gi}		$i = 1\text{--}4$
Pinion					θ_{pi}		$i = 1\text{--}4$
Rotor					θ_{ri}		$i = 1\text{--}4$

2.2. Electric Model

Due to the complex motion relationship between stator and rotor, the mathematical model of a three-phase induction motor is a strongly coupled nonlinear multivariable system [40]. In order to facilitate the establishment of the motor control system, the linear and decoupled mathematical model is often obtained through coordinate transformation [41–46]. In addition, the direct torque control method is often used in large mechanical systems, such as rail vehicles [47,48], because of its natural advantages of quick response and convenience.

2.2.1. Mathematical Model of the Traction Motor

The simplified model of the motor has the following assumptions. First, the three-phase winding is symmetrical. The magnetic force is sinusoidally distributed along the circumference of the air gap. Second, the effect of magnetic saturation, core loss, and temperature and frequency influence on motor resistance is ignored based on the mathematical model theory of asynchronous motor [49]. The mathematical expression of coordinate transformation is usually expressed in matrix equation form as Equation (1):

$$Y = AX \tag{1}$$

where A is the transformation matrix, X is the original vector before transformation, and Y is the new vector after transformation.

The above matrix transformation can transform the current matrix in the three static coordinate systems into a new matrix in another coordinate system. Simultaneously, the process satisfies the principle of invariable power before and after transformation.

As shown in Figure 3 below, in the three-phase stationary coordinate system, the A-axis is recombined with the α-axis in the two-phase coordinate system. It is assumed that the adequate number of turns in each phase of the stator winding is N, and the magnetic emf waveform is sinusoidal. The component algebras of the three-phase magnetomotive force on the α and β axes are equal to those of the two-phase magnetomotive force. Equation (2) expresses the relationship as below.

$$\begin{cases} N_2 i_{s\alpha} = N_3 i_A + N_3 i_B \cos \frac{2\pi}{3} + N_3 i_C \cos \frac{4\pi}{3} \\ N_2 i_{s\beta} = 0 + N_3 i_B \sin \frac{2\pi}{3} + N_3 i_C \sin \frac{4\pi}{3} \end{cases} \tag{2}$$

Equation (3) is the transformation matrix from a three-phase stationary coordinate to a two-phase synchronous coordinate.

$$\begin{bmatrix} i_\alpha \\ i_\beta \end{bmatrix} = \sqrt{\frac{2}{3}} \begin{bmatrix} 1 & -\frac{1}{2} & -\frac{1}{2} \\ 0 & \frac{\sqrt{3}}{2} & -\frac{\sqrt{3}}{2} \end{bmatrix} \begin{bmatrix} i_A \\ i_B \\ i_C \end{bmatrix} \tag{3}$$

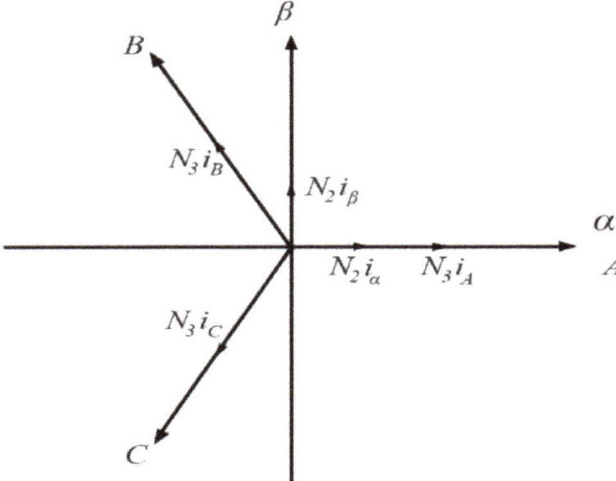

Figure 3. Schematic diagram of space vector of magnetomotive force.

Based on the above coordinate system transformation, the mathematical model of asynchronous motor is established in any two-phase coordinate system, and Equations (4) and (5) can express the magnetic flux and voltage equations, respectively:

$$
\begin{bmatrix} \varphi_{sd} \\ \varphi_{sq} \\ \varphi_{rd} \\ \varphi_{rq} \end{bmatrix} = \begin{bmatrix} L_s & 0 & L_m & 0 \\ 0 & L_s & 0 & L_m \\ L_m & 0 & L_r & 0 \\ 0 & L_m & 0 & L_r \end{bmatrix} \begin{bmatrix} i_{sd} \\ i_{sq} \\ i_{rd} \\ i_{rq} \end{bmatrix}
\tag{4}
$$

where the symbols φ_{sd} and φ_{sq} are the components of stator flux linkage in the two-phase coordinate system, respectively, and φ_{rd} and φ_{rq} are the components of rotor flux in the two-phase coordinate system; i_{sd}, i_{sq}, i_{rd}, and i_{rq} are the components of stator current and rotor current in the two-phase coordinate system, respectively; L_s and L_r are the self-inductance of stator equivalent two-phase windings and rotor equivalent two-phase windings in the two-phase coordinate system, respectively, and L_m is the mutual inductance between coaxial equivalent windings of stator and rotor in the two-phase coordinate system.

$$
\begin{bmatrix} u_{sd} \\ u_{sq} \\ u_{rd} \\ u_{rq} \end{bmatrix} = \begin{bmatrix} R_s & 0 & 0 & 0 \\ 0 & R_s & 0 & 0 \\ 0 & 0 & R_r & 0 \\ 0 & 0 & 0 & R_r \end{bmatrix} \begin{bmatrix} i_{sd} \\ i_{sq} \\ i_{rd} \\ i_{rq} \end{bmatrix} + \begin{bmatrix} L_s p & 0 & L_m p & 0 \\ 0 & L_s p & 0 & L_m p \\ L_m p & 0 & L_r p & 0 \\ 0 & L_m p & 0 & L_r p \end{bmatrix} \begin{bmatrix} i_{sd} \\ i_{sq} \\ i_{rd} \\ i_{rq} \end{bmatrix}
$$
$$
+ \begin{bmatrix} 0 & -\omega_{dqs} & 0 & 0 \\ \omega_{dqs} & 0 & 0 & 0 \\ 0 & 0 & 0 & -\omega_{dqr} \\ 0 & 0 & \omega_{dqr} & 0 \end{bmatrix} \begin{bmatrix} \varphi_{sd} \\ \varphi_{sq} \\ \varphi_{rd} \\ \varphi_{rq} \end{bmatrix}
\tag{5}
$$

where the symbols u_{sd}, u_{sq}, u_{rd}, and u_{rq} are components of stator voltage and rotor voltage in the two-phase coordinate system, respectively; R_s and R_r represent the stator resistance and rotor resistance, respectively; p is a differential operator, ω_{dqs} is the angular velocities relative to the stator, and ω_{dqr} is the angular velocity relative to the rotor in the two-phase coordinate system.

According to the kinematics theory [50], the output torque equation of the motor can be expressed by Equation (6), and the motion equation of the motor can be expressed by Equation (7).

$$
T_e = n_p L_m \left(i_{sq} i_{rd} - i_{sd} i_{rq} \right)
\tag{6}
$$

$$
T_e = T_m + \frac{J}{n_p} \frac{d\omega_r}{dt} + \frac{D}{n_p} \omega_r
\tag{7}
$$

where T_e is the electromagnetic torque of the motor, n_p is polar logarithm, T_m is the load torque, J is the moment of inertia of the electromechanical system, and D is the damping coefficient.

The research in this paper uses a voltage inverter, and Figure 4 shows the schematic diagram of main circuit of traction inverter. The corresponding power grid obtains the dc voltage through the transformer and the rectifier circuit. It outputs the voltage value by switching the inverter to control the motor. If the three-phase load is connected to different phases, the switching state of the inverter is 1. Otherwise, when the three-phase load is connected to the same phase, the switching state of the inverter is 0. Figure 5 is the voltage vector schematic diagram of the inverter. So, there are eight combinations of switching states of inverters, which are as follows: U_0 (000), U_1 (100), U_2 (110), U_3 (010), U_4 (011), U_5 (001), U_6 (101), and U_7 (111).

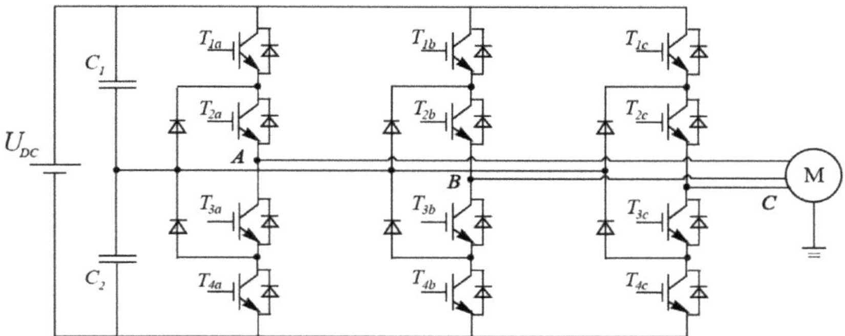

Figure 4. The main circuit of the thre- level voltage-type traction inverter.

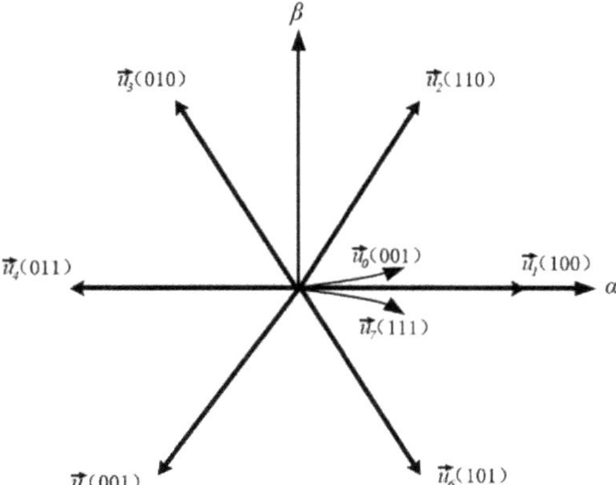

Figure 5. Schematic diagram of inverter voltage vector.

The eight voltage state-space vectors of the inverter form eight discontinuous voltage space vectors. The angle between the two adjacent vectors in the six non-zero voltage vectors is 60°. The counterclockwise rotation order of the vectors is as follows: $U_1 \rightarrow U_2 \rightarrow U_3 \rightarrow U_4 \rightarrow U_5 \rightarrow U_6$. The stator voltage u_s in any of the switching states can be expressed as a vector in the two-phase coordinate system.

$$u_s(t) = \frac{2}{3}\left[u_A + u_B e^{j2\pi/3} + u_C e^{j4\pi/3}\right] \tag{8}$$

where u_A, u_B, and u_C are phase voltages of three-phase stator load, respectively.

When the traction motor is connected to the non-sinusoidal power, the time harmonic magneto-motive force will be generated in the air gap of the motor, which will generate additional harmonic torque. When the air gap harmonic flux and harmonic rotor current have the same order, their interaction will produce a stable harmonic torque, and when the times of harmonic flux and harmonic rotor current are different, their interaction will produce a vibration harmonic torque. If the fundamental and harmonic waves in the air gap generate n rotating magnetic fields, there will be (n − 1) stable harmonic torques, and Equation (9) is the calculation formula of kth harmonic torque:

$$T_k = \pm \frac{mn_p}{2\pi f_1} I_{2k}^2 \frac{R_{rk}}{(k \pm 1)} \tag{9}$$

where m is the number of motor stators, n_p is polar logarithm, f_1 is the input fundamental wave voltage frequency of the motor stator, I_{2k} is the calculated value of rotor current under kth harmonics, and R_{rk} is the rotor resistance calculated to the stator side under kth harmonics.

If the fundamental and harmonic waves in the air gap generate n rotating magnetic fields, there will be (n^2 − n) vibration harmonic torques. Equation (10) gives the fifth harmonic vibration torque and Equation (11) gives the 7th harmonic vibration torque.

$$T_{5-1} = -\frac{3n_p}{2\pi f_1} I_{25} E_2 \cos(6\omega t - \phi_2) = \frac{3n_p}{2\pi f_1} I_{25} E_2 \cos(6\omega t + \pi - \phi_2) \tag{10}$$

$$T_{7-1} = \frac{3n_p}{2\pi f_1} I_{27} E_2 \cos(6\omega t - \phi_2) \tag{11}$$

where E_2 is the calculated value of the emf of fundamental rotor, and ϕ_2 is the phase difference between current and electromotive force when $\omega t = 0$.

When the train is in traction operation, the pantograph transforms the AC power of cate nary into DC power through high-voltage electrical equipment, a traction transformer and traction converter, and then the traction inverter drives the traction motor by outputting three-phase AC power. During the braking deceleration, the traction inverter is controlled to make the traction motor in the state of power generation, and the traction inverter feeds the three-phase ac power output of the traction motor back to the catenary through the rectification link. As shown in Table 3, we can use A and B to represent the power absorbed and fed back by the traction inverter from the power grid, respectively:

$$\begin{cases} P_F = \frac{16P_M}{\eta_g \eta_M \eta_i} + S_{aux} \cos \varphi_{aux} \\ P_B = P_M \eta_g \eta_M \eta_i - S_{aux} \cos \varphi_{aux} \end{cases} \tag{12}$$

where P_M is the output power of traction motor.

Table 3. Parameter table of traction and the auxiliary inverter.

Parameter	Value
Traction drive system efficiency (η_g)	0.95
Conversion efficiency of power electronic equipment (η_i)	0.96
Traction motor efficiency (η_M)	0.94
Capacity of auxiliary inverter (S_{aux})	4 × 394 KW
Auxiliary inverter power factor ($\cos\varphi_{aux}$)	0.87

2.2.2. Asynchronous Motor Control Strategy

This paper uses the direct torque control method to realize the effective control of electromagnetic torque of asynchronous traction motor. The direct torque control method uses magnetic flux and electromagnetic torque as control variables. A discrete inverter vector controls the stator flux vector trajectory. The control method is fast and straightforward, which significantly improves the dynamic response ability of the system.

As shown in Figure 6, the basic flow of the direct torque control method is as follows: The voltage and current signals u_a, u_b, u_c, i_a, i_b, and i_c are sent by the inverter to the asynchronous traction motor are obtained by the measurement module. Then, the components u_α, u_β, i_α, and i_β of the stator three-phase voltage and current signals after coordinate transformation are obtained by using the Clark transform method [51]. After that, the stator flux and actual torque model are then obtained according to voltage and current components.

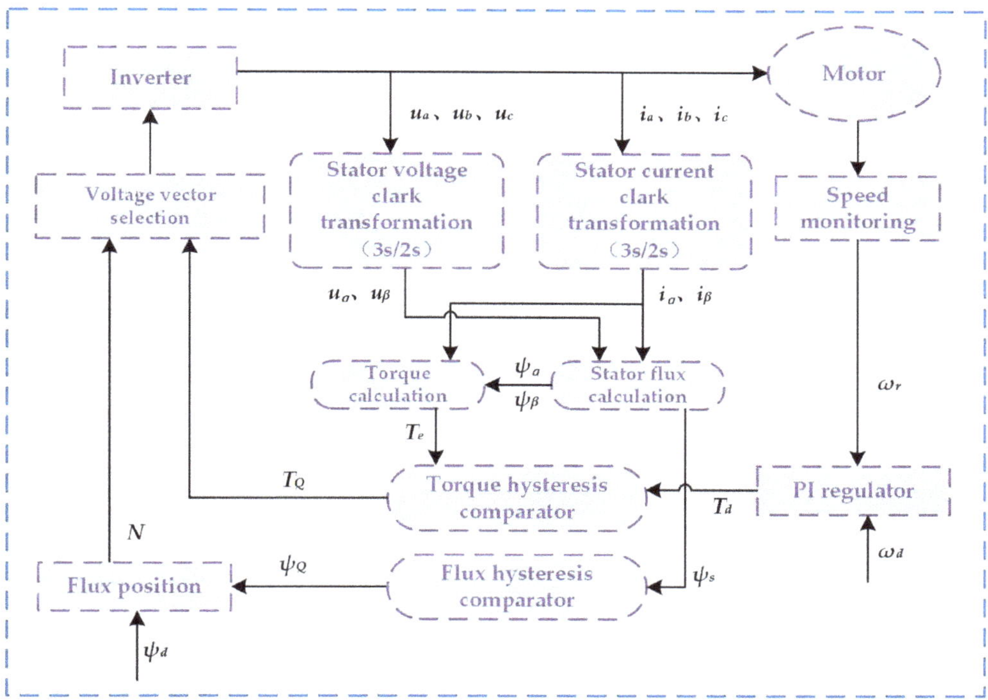

Figure 6. Principle block diagram of direct torque control for traction motor.

The direct torque control theory [52] shows that the motor stator flux vector and the actual motor torque can be expressed as Equations (13) and (14).

$$\psi_\alpha = \int (u_\alpha - R_s i_\alpha)dt$$
$$\psi_\beta = \int (u_\beta - R_s i_\beta)dt \tag{13}$$

$$T_e = n_p(\psi_\alpha i_\beta - \psi_\beta i_\alpha) \tag{14}$$

where the symbols ψ_α and ψ_β represent the components of the magnetic vector in a two-phase coordinate system.

Furthermore, the rotational angular velocity ω_r of the rotor measured by the speed sensor and the given angular velocity ω_d are calculated by P.I. adjustment. As a result, the motor's reference torque value T_e^* is obtained. Then, the reference torque value T_e^* and the actual calculated torque value T_e are used to obtain the torque switch signal $T_{\cdot Q \cdot}$ through the hysteresis comparator. Similarly, the flux amplitude ψ_s obtained by the flux calculation model and the given flux value ψ_s^* can be used to obtain the flux switch signal ψ_Q by the hysteresis comparator. Meanwhile, the stator flux calculation model determines the interval value N of flux through the S-function. On this basis, the voltage selector switch signal unit is combined with flux interval value N, flow switch signal ψ_Q, and torque switch signal T_Q. The corresponding voltage switch vector is obtained by the program written by S-function to control the inverter output controllable three-phase A.C. signal.

2.3. Electromechanical Coupling Model

The resistance is an unavoidable external force when the train runs on the line. The running resistance of a high-speed train mainly includes basic resistance and additional resistance. The basic resistance refers to the resistance existing in any operating condition of the train, and the additional resistance refers to the resistance generated by the train in the ramp, curve, tunnel, and other individual working conditions. Therefore, in this paper, the model only considers the influence of the basic resistance, and according to the train traction calculation theory [53,54], its expression is:

$$\omega = a + bv + cv^2 \tag{15}$$

where ω is the basic resistance per unit mass, and its unit is N/t; a is the rolling resistance coefficient, which is 8.63 N; b is the swing vibration resistance coefficient, which is 0.07295 N·h/t; and c is the air resistance coefficient of train operation, which is 0.00112 N·h/t.

The electromechanical coupling model of a quarter traction drive system is shown in Figure 7. This section describes the coupling process of vehicle mechanical and electrical structures. The co-simulation is between the vehicle dynamic model based on Simpack and the traction motor model based on Simulink by third-party interfaces. The SIMAT interface is realized in the Simpack function module. Firstly, to make the dynamic model normally run in Simulink software, it is necessary to package the dynamic model of the vehicle and set the input and output of the dynamic model. Then, based on the direct torque control method, the torque output of the traction motor is taken as the input of the dynamic model. The real-time angular velocity of the rotor of the vehicle model is the output of the dynamic model and the input value of the traction motor model. Subsequently, the traction motor compares the angular velocity value output by the dynamic model with the given angular velocity value to make a rapid response to control the motor torque output. In addition, the input of the dynamic model uses the basic resistance value calculated from the current running speed of the vehicle model.

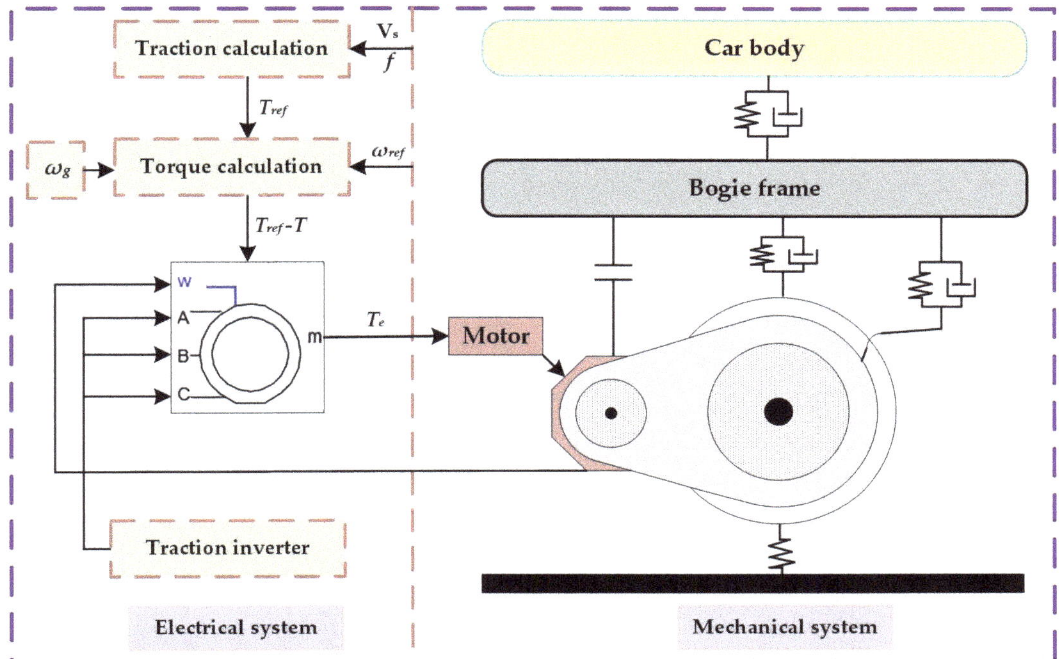

Figure 7. Electromechanical coupling model of a quarter drive subsystem.

3. Numerical Simulation under Variable Conditions

This section describes the specific variable conditions of the simulation. Additionally, the vehicle model simulation in this paper ignores the influence of rail irregularity. First, the vehicle starts to accelerate traction from the initial speed of 0 km/h. During the traction process, the maximum speed of the motor rotor is set to 153 r/s, and the vehicle speed is about 100 km/h at this time. The motor then controls the vehicle to enter the braking process when it travels at a constant speed of 100 km/h for 50 s. Next, in the braking process, the motor speed is set to 60 r/s, and the vehicle running speed is about 40 km/h at this time. Finally, when the rotor speed tends to stabilize, the vehicle runs at a constant speed of 40 km/h for 100 s. Figure 8 below indicates the variation of the motor output torque and vehicle speed.

In Figure 8, the output torque of the traction motor is about 2500 N·m in the process of traction acceleration and −2500 N·m in the braking process. However, the motor's output torque fluctuates around 0 N·m in the process of constant speed, and the fluctuation range is about ±150 N·m. As shown from Figure 8, under the traction acceleration process, the traction motor's rotor speed accelerates to 153 r/s, and the vehicle accelerates to 100 km/h at the time of 29.3 s. The vehicle then runs at a constant speed for 50 s. After that, the motor outputs the braking torque. Finally, the vehicle enters a stable running state at around 63 s with a motor speed of about 60 r/s (40 km/h for vehicle speed).

In Figure 9, the stator three-phase current amplitude of the traction motor is prominent in the traction and braking process but small with the constant speed. During the transition from traction acceleration/braking deceleration to constant speed, the current changes relatively gently. It can be seen from Figure 10a that the motor stator current of the vehicle is considerable at the moment of starting. The amplitude reaches about 1400 A. Measures should be taken to avoid damage to the motor caused by the excessive starting current.

As shown in Figure 10c, when the vehicle is braking, the three-phase current of the motor stator changes abruptly.

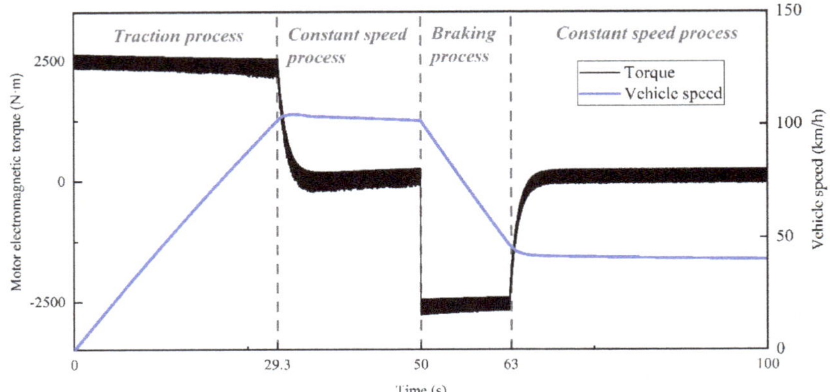

Figure 8. The variation of motor torque and vehicle speed with time.

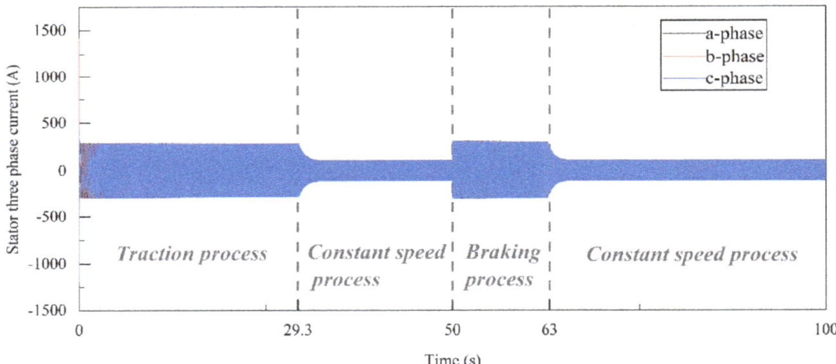

Figure 9. Time histories of the three-phase current of the motor stator.

As is shown in Figure 11a, after the vehicle starts, with the increase of the vehicle speed, the RMS of the stator three-phase current of the traction motor decreases gradually from about 200 A in the traction process. Figure 11c shows the braking process, the RMS of the three-phase current of the motor stator is greater than that of the traction process, and the RMS of the current is about 206 A. With the decrease of the vehicle speed, the RMS of the current gradually decreases. In a constant speed process at 100 km/h and 40 km/h, the RMS values vary between 71 A and 73 A.

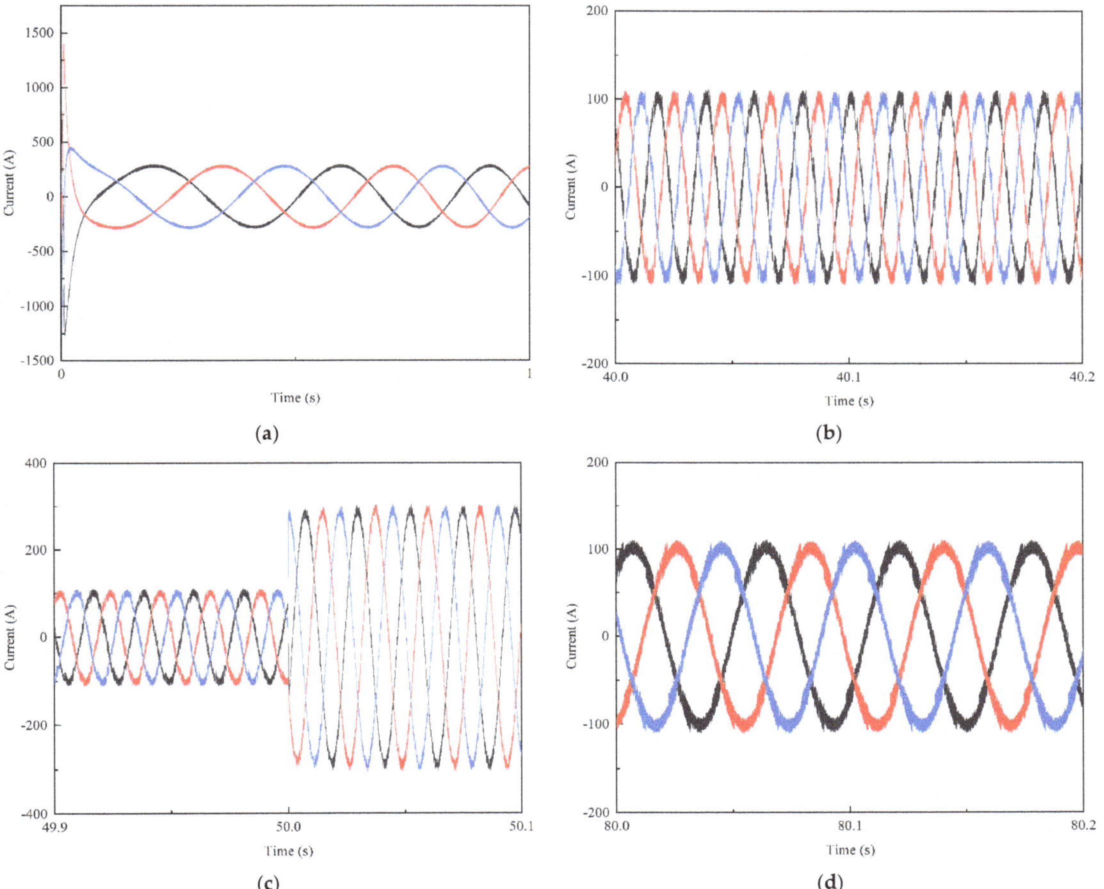

Figure 10. The stator current in (**a**) the traction process, (**b**) the constant speed process at 100 km/h, (**c**) the braking process, and (**d**) the constant speed process at 40 km/h.

In Figures 12–14, the time-frequency analysis of the stator three-phase current indicates that the stator three-phase current frequency gradually rises with the vehicle speed increases. Conversely, the stator three-phase current frequency progressively decreases during the braking process. The theory of the relationship between gear speed and meshing frequency provides the formulas of gear meshing frequency (Equation (16)) and gear shaft rotation frequency (Equation (17)).

$$f_c = \frac{V \cdot N}{\pi \cdot D} \tag{16}$$

$$f_r = \frac{f_c}{Z} \tag{17}$$

where the symbol f_c is the meshing frequency of the gear pair, V, N, D, and Z are the vehicle speed, gear teeth, wheel diameter, and pinion teeth, respectively, and f_r is the rotation frequency of the pinion shaft.

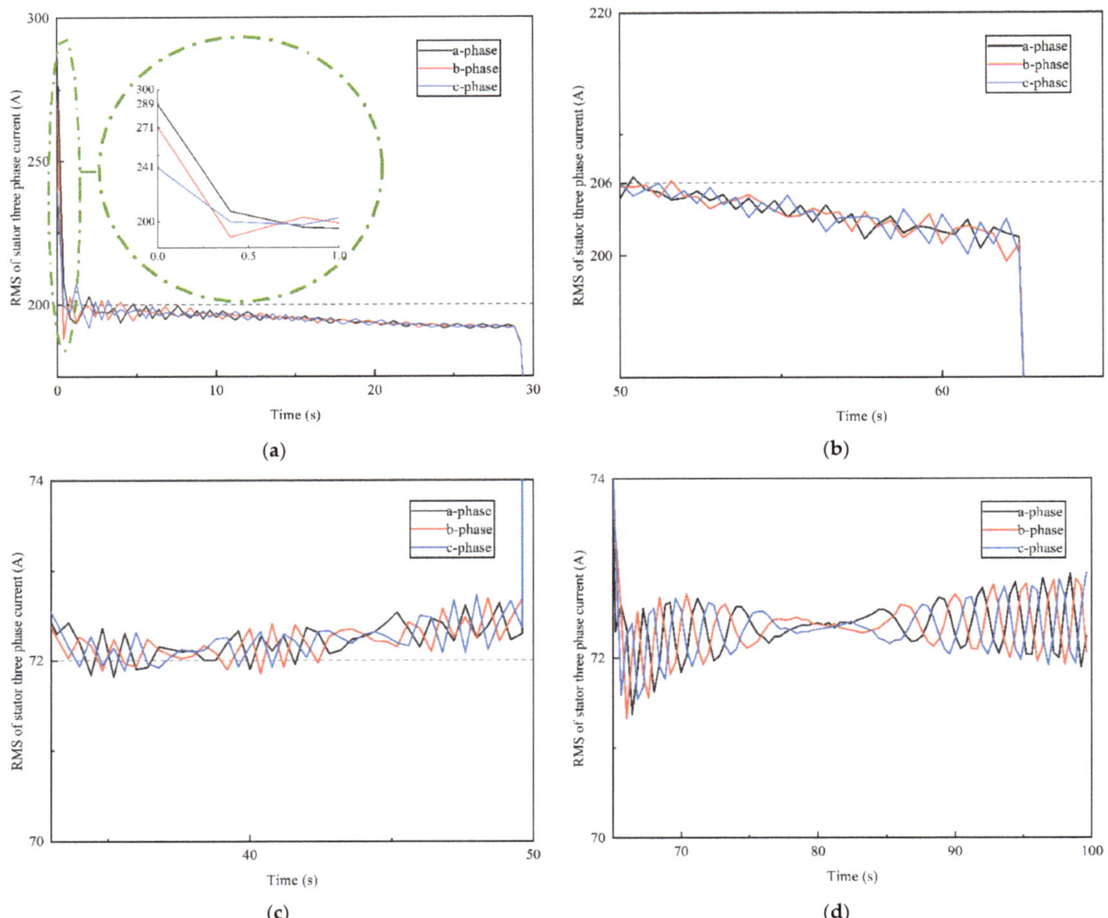

Figure 11. The RMS value of the stator current in (**a**) the traction process, (**b**) the constant speed process at 100 km/h, (**c**) the braking process, and (**d**) the constant speed process at 40 km/h.

When the speed reaches 100 km/h, the stator current frequency is about 49 Hz, and the rotation frequency of the pinion shaft is 24.5 Hz. Furthermore, the stator current frequency is twice the rotational frequency of the pinion shaft. Like the braking process, the stator three-phase current frequency is around 19.5 Hz when the vehicle speed is 40 km/h. Additionally, the rotation frequency of the pinion shaft is 9.8 Hz. It is worth mentioning that the gear meshing frequency f_c can also be displayed in the current time-frequency diagram. However, its frequency multiplication is relatively less evident due to the significant meshing frequency of the gear. In addition, harmonic components such as five times and seven times the stator three-phase current can be seen in the time-frequency analysis, mainly due to the control mode of the motor.

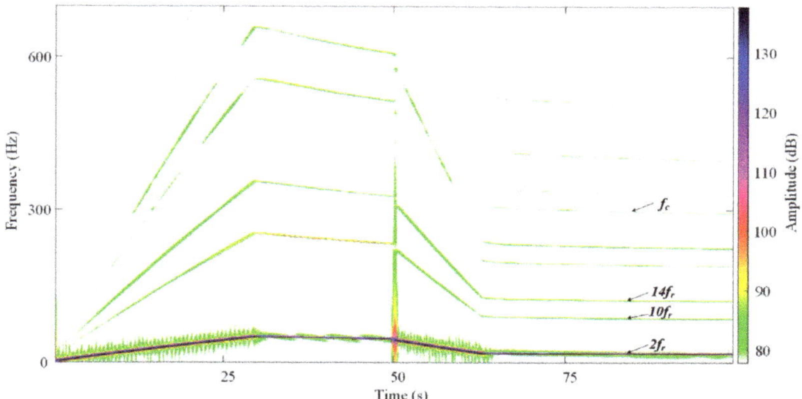

Figure 12. Time-frequency characteristics of stator three-phase current in a-phase.

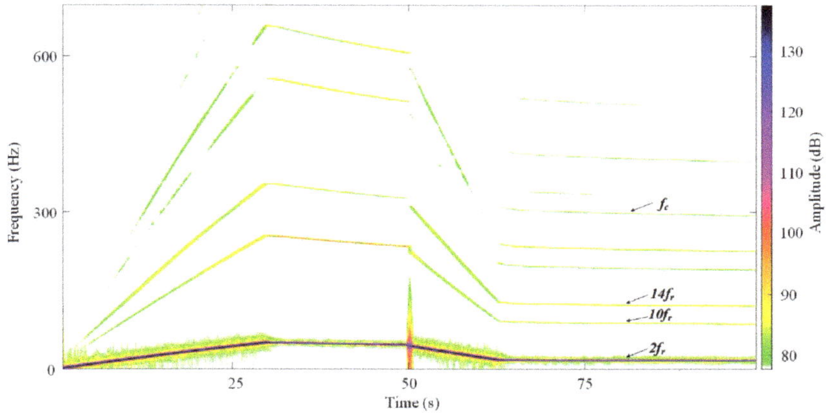

Figure 13. Time-frequency characteristics of stator three-phase current in b-phase.

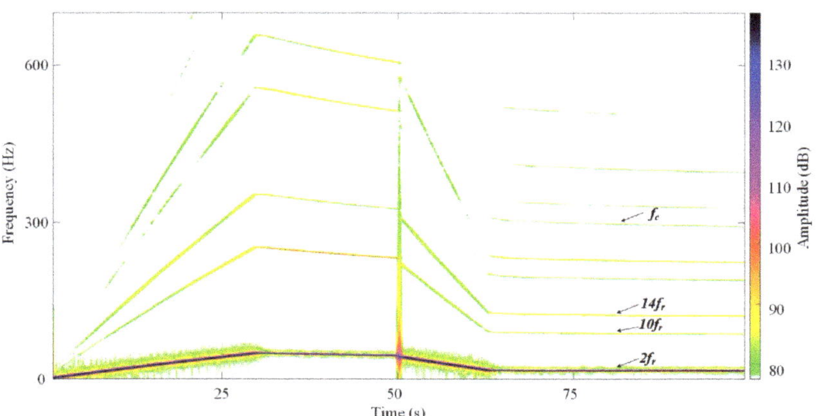

Figure 14. Time-frequency characteristics of stator three-phase current in c-phase.

It can be seen from Figure 15 that, at the moment of vehicle starting, the rotor current also has a significant mutation reaching 1250 A, which can be clearly seen in Figure 16a. Compared with the stator three-phase current, the current amplitude decreases significantly from the traction or braking process to the uniform speed process. The rotor three-phase current is relatively stable at a specific value in the uniform speed process. However, the rotor current has a noticeable abrupt change from constant speed to braking. It can be seen from Figure 17a that, in the process of vehicle traction acceleration, the RMS of rotor three-phase current is about 185 A. With the increase of vehicle speed, the RMS value of rotor current tends to decrease. Figure 17c shows that the RMS of rotor current in the braking process is larger than that in the traction process, which is about 190 A and tends to decrease with the vehicle speed. When the vehicle runs at a constant speed of 100 km/h and 40 km/h, the RMS value of the rotor current varies from 6 A to 13 A, which is relatively stable. The rotor current has a more expansive change period and a more minor frequency than the stator current.

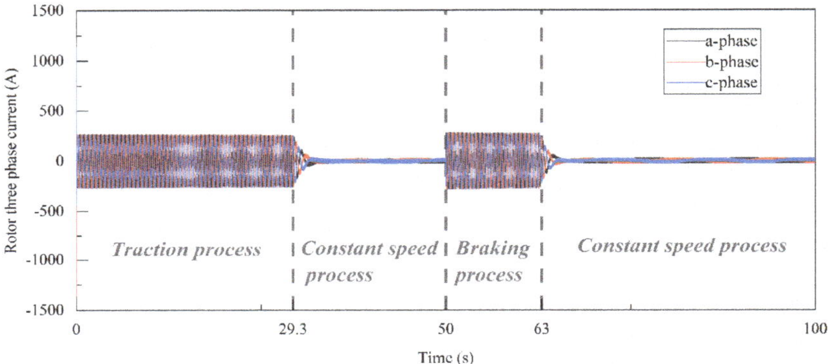

Figure 15. Time histories of the three-phase current of the motor rotor.

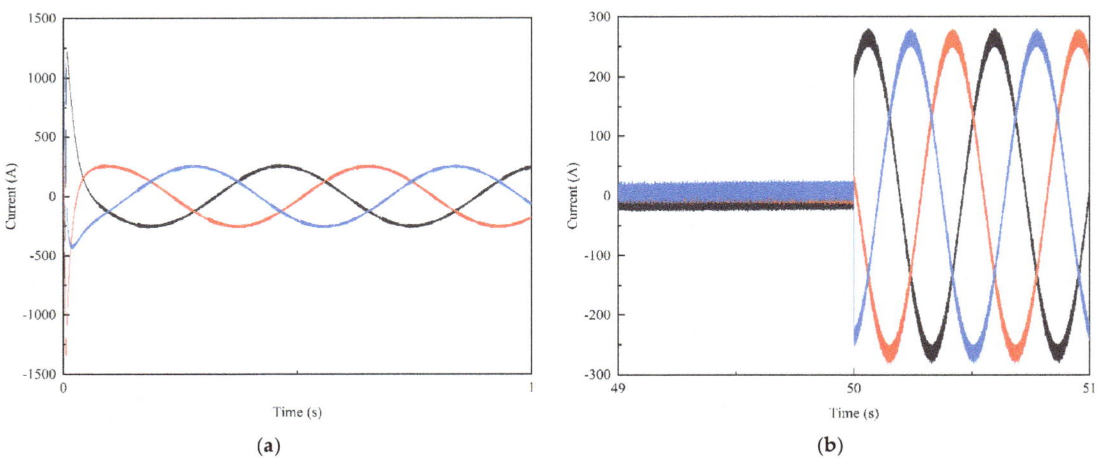

(**a**)

(**b**)

Figure 16. The variation of rotor current with time at the beginning of (**a**) the traction process and (**b**) the braking process.

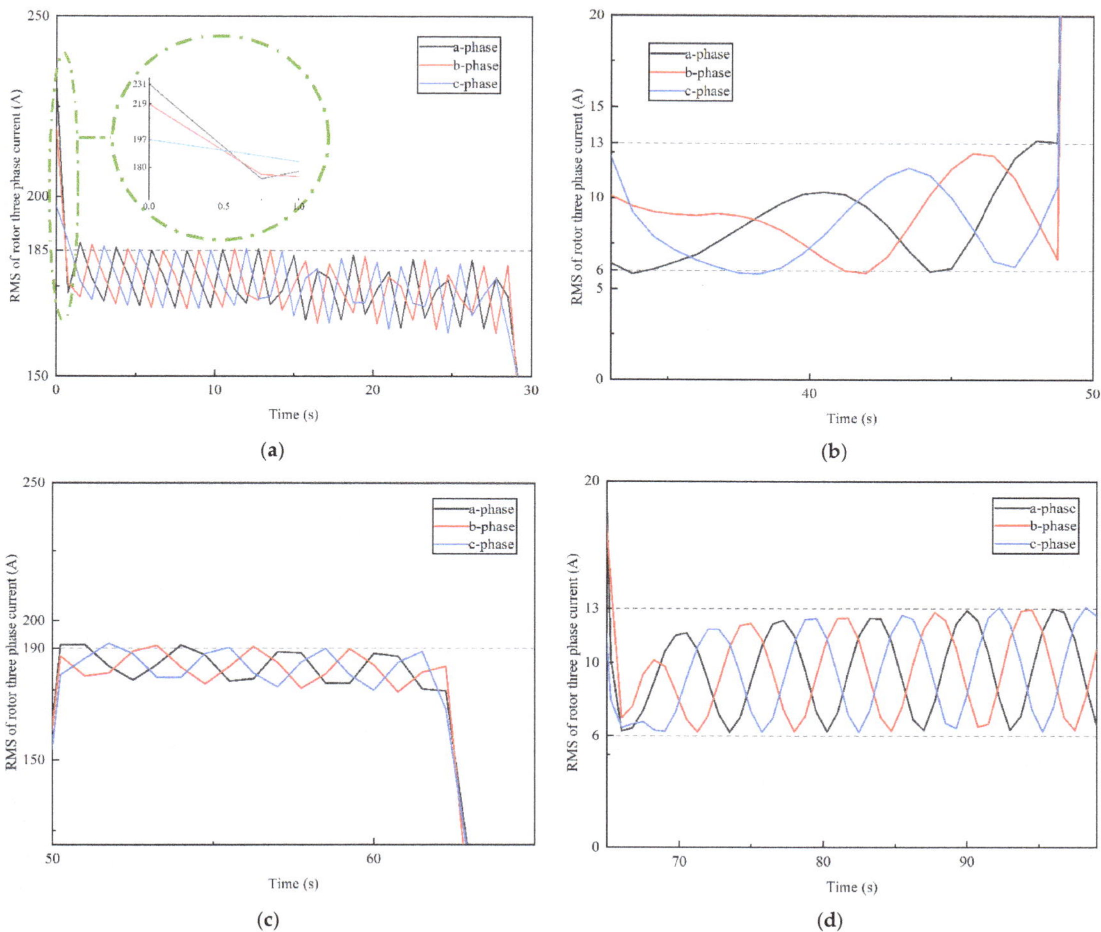

Figure 17. The RMS value of the rotor current in (**a**) the traction process, (**b**) the constant speed process at 100 km/h, (**c**) the braking process, and (**d**) the constant process at 40 km/h.

4. Conclusions

This paper established an electromechanical coupling model combined with the traction control and vehicle models. The co-simulation is under traction, constant speed, and braking conditions. Additionally, this research reveals the dynamic characteristics of rotor and stator current during vehicle operation through numerical research. Overall, the conclusions are given as follows.

When the vehicle starts, the current of traction motor stator and rotor have a significant mutation, so measures should be taken to protect the motor. The stator current amplitude reaches 1400 A, and the amplitude of the rotor current reaches 1250 A. The stator current and rotor current gradually decrease with the vehicle speed increase in the traction process. Similarly, the stator current and rotor current decrease gradually with the vehicle speed decrease in the braking process. Both the stator current and rotor current maintain a stable amplitude in the constant speed process.

In order to analyze the change trend of current amplitude of traction motor under the transient condition of variable speed and load, the RMS value is used to describe the change of current amplitude. Similarly, short-time Fourier transform (STFT) is used to analyze the time-frequency characteristics of the current. In the traction and braking process, the RMS value of the stator three-phase current is smaller than that of the braking process. Moreover, the RMS value tends to decrease with the increase or decrease of the vehicle speed during traction and braking. In the traction process, the RMS value of stator current decreases gradually from about 200 A, and while braking, the RMS value of stator current decreases gradually from about 206 A. Moreover, in the constant speed process of 100 km/h and 40 km/h, the RMS value varies between 71 A and 73 A, respectively. The high harmonic of driving shaft rotation frequency and the frequency of gear meshing can be seen in the stator current spectrum. At the vehicle speed of 100 km/h, the stator current frequency (about 49 Hz) is twice as high as the pinion shaft rotation frequency (about 24.5 Hz). The same pattern is observed at a speed of 40 km/h. Except for the rotational frequencies at two constant speeds, these higher harmonics are evident and obscure during the traction and braking process.

In the traction process, the RMS value of the rotor three-phase current is smaller than that of braking. Therefore, during the traction process and braking process, the RMS value of the current all tends to decrease gradually. In the traction process, the RMS value of the rotor current decreases gradually from about 185 A, and while braking, the RMS value of rotor current decreases gradually from about 190 A. The RMS value of the rotor current fluctuates stably in the range of 6 A to 13 A during constant speed. However, it tends to decrease during the traction and braking process.

Author Contributions: Conceptualization, K.Z., J.Y., C.L., J.W. and D.Y.; methodology, K.Z., J.Y. and J.W.; software, K.Z. and J.W.; validation, K.Z., J.Y., C.L., J.W. and D.Y.; formal analysis, K.Z., J.Y. and J.W.; investigation, K.Z. and J.Y.; resources, J.Y., C.L., J.W. and D.Y.; data curation, K.Z., J.Y. and J.W.; writing—original draft preparation, K.Z.; writing—review and editing, K.Z.; visualization, K.Z.; supervision, J.Y., C.L., J.W. and D.Y.; project administration, J.Y., C.L., J.W. and D.Y.; funding acquisition, J.Y., C.L., J.W. and D.Y. All authors have read and agreed to the published version of the manuscript.

Funding: This paper was supported by the National Natural Science Foundation of China (grant number 51975038), the Beijing Natural Science Foundation (grant number 3214042, KZ202010016025, L191005), the Beijing Postdoctoral Research Foundation (grant number 2021-zz-114), the Open Research Fund Program of Beijing Key Laboratory of Performance Guarantee on Urban Rail Transit Vehicles (grant number PGU2020K001), the Fundamental Research Funds for the Beijing Universities (grant number X21049), and the Development of High-Level Teachers in Beijing Municipal Universities (grant number CIT&TCD201904062).

Institutional Review Board Statement: Not applicable.

Informed Consent Statement: Not applicable.

Data Availability Statement: Not applicable.

Conflicts of Interest: The authors declare no conflict of interest.

References

1. Feng, J.H.; Wang, J.; Li, J.H. Integrated Simulation Platform of High-Speed Train Traction Drive System. *Tiedao Xuebao/J. China Railw. Soc.* **2012**, *34*, 21–26.
2. Chen, Z.G.; Shao, Y.M.; Lim, T.C. Non-Linear Dynamic Simulation of Gear Response under the Idling Condition. *Int. J. Autom. Technol.* **2012**, *13*, 541–552. [CrossRef]
3. Shantarenko, S.; Kuznetsov, V.; Ponomarev, E.; Vaganov, A.; Evseev, A. Influence of Process Parameters on Dynamics of Traction Motor Armature. *Transp. Res. Proced.* **2021**, *54*, 961–971. [CrossRef]
4. Wang, Z.; Mei, G.; Zhang, W.; Cheng, Y.; Zou, H.; Huang, G. Effects of Polygonal Wear of Wheels on the Dynamic Performance of the Gearbox Housing of a High-Speed Train. *Proc. Inst. Mech. Eng. Part F J. Rail Rapid Transit* **2018**, *232*, 095440971775299. [CrossRef]
5. Wang, Z.; Cheng, Y.; Mei, G.; Zhang, W.; Huang, G.; Yin, Z. Torsional Vibration Analysis of the Gear Transmission System of High-Speed Trains with Wheel Defects. *Proc. Inst. Mech. Eng. Part F J. Rail Rapid Transit* **2019**, *234*, 095440971983379. [CrossRef]
6. Henao, H.; Kia, S.H.; Capolino, G.A. Torsional-Vibration Assessment and Gear-Fault Diagnosis in Railway Traction System. *IEEE Trans. Ind. Electron.* **2011**, *58*, 1707–1717. [CrossRef]

7.   Garg, V.K.; Dukkipati, R.V. *Dynamics of Railway Vehicle Systems*; Academic Press: Cambridge, MA, USA, 1984; pp. 245–247.
8.   Zhai, W. The Vertical Model of Vehicle-Track System and Its Coupling Dynamics. *J. China Railw. Soc.* **1992**, *14*, 10–21.
9.   Ren, Z.S.; Zhai, W.M.; Wang, Q.C. Study on Lateral Dynamic Characteristics of Vehicle Turnout System. *J. China Railw. Soc.* **2000**, *22*, 28–33. [CrossRef]
10. Nielsen, J.; Igeland, A. Vertical Dynamic Interaction between Train and Track Influence of Wheel and Track Imperfections. *J. Sound Vib.* **1995**, *187*, 825–839. [CrossRef]
11. Zhai, W.M.; Wang, K.Y. Lateral Interactions of Trains and Tracks on Small-Radius Curves: Simulation and Experiment. *Veh. Syst. Dyn.* **2006**, *44* (Suppl. S1), 520–530. [CrossRef]
12. Wang, J.; Yang, J.; Zhao, Y.; Bai, Y.; He, Y. Nonsmooth Dynamics of a Gear-Wheelset System of Railway Vehicles under Traction/Braking Conditions. *J. Comput. Nonlinear Dyn.* **2020**, *15*, 081003. [CrossRef]
13. Wang, J.; Yang, J.; Li, Q. Quasi-Static Analysis of the Nonlinear Behavior of a Railway Vehicle Gear System Considering Time-Varying and Stochastic Excitation. *Nonlinear Dyn.* **2018**, *93*, 463–485. [CrossRef]
14. Huang, G.H.; Zhou, N.; Zhang, W.H. Effect of Internal Dynamic Excitation of the Traction System on the Dynamic Behavior of a High-Speed Train. *Proc. Inst. Mech. Eng. Part F J. Rail Rapid Transit* **2016**, *230*, 0954409715617787. [CrossRef]
15. Zhang, W.H.; Shen, Z.Y.; Zeng, J. Study on Dynamics of Coupled Systems in High-Speed Trains. *Veh. Syst. Dyn. Int. J. Veh. Mech. Mobil.* **2013**, *51*, 966–1016. [CrossRef]
16. Yao, Y.; Zhang, H.J.; Luo, S.H. Analysis on Resonance of Locomotive Drive System under Wheel-Rail Saturated Adhesion. *J. China Railw. Soc.* **2011**, *33*, 16–22.
17. Yao, Y.; Zhang, H.J.; Luo, S.H. An Analysis of Resonance Effects in Locomotive Drive Systems Experiencing Wheel/Rail Saturation Adhesion. *Proc. Inst. Mech. Eng. Part F J. Rail Rapid Transit* **2014**, *228*, 4–15. [CrossRef]
18. Farshidianfar, A.; Saghafi, A. Global Bifurcation and Chaos Analysis in Nonlinear Vibration of Spur Gear Systems. *Nonlinear Dyn.* **2014**, *75*, 783–806. [CrossRef]
19. Chen, Z.; Zhou, Z.; Zhai, W.; Wang, K. Improved Analytical Calculation Model of Spur Gear Mesh Excitations with Tooth Profile Deviations. *Mech. Mach. Theory* **2020**, *149*, 103838. [CrossRef]
20. Chen, Z.; Zhai, W.; Wang, K. Vibration Feature Evolution of Locomotive with Tooth Root Crack Propagation of Gear Transmission System. *Mech. Syst. Signal Process.* **2019**, *115*, 29–44. [CrossRef]
21. Chen, Z.; Zhu, Z.; Shao, Y. Fault Feature Analysis of Planetary Gear System with Tooth Root Crack and Flexible Ring Gear Rim. *Eng. Fail. Anal.* **2015**, *49*, 92–103. [CrossRef]
22. Chen, Z.; Zhai, W.; Wang, K. Dynamic Investigation of a Locomotive with Effect of Gear Transmissions under Tractive Conditions. *J. Sound Vib.* **2017**, *408*, 220–233. [CrossRef]
23. Wang, J.; Yang, J.; Lin, Y.; He, Y. Analytical Investigation of Profile Shifts on the Mesh Stiffness and Dynamic Characteristics of Spur Gears. *Mech. Mach. Theory* **2022**, *167*, 104529. [CrossRef]
24. Wang, J.; Yang, J.; Bai, Y.; Zhao, Y.; Yao, D. A Comparative Study of the Vibration Characteristics of Railway Vehicle Axlebox Bearings with Inner/Outer Race Faults. *Proc. Inst. Mech. Eng. Part F J. Rail Rapid Transit* **2020**, *235*, 1–13. [CrossRef]
25. Liu, P.F.; Zhai, W.M.; Wang, K.Y. Establishment and Verification of Three-Dimensional Dynamic Model for Heavy-Haul Train–Track Coupled System. *Veh. Syst. Dyn.* **2016**, *54*, 1511–1537. [CrossRef]
26. Liu, Y.; Chen, Z.; Tang, L.; Zhai, W. Skidding Dynamic Performance of Rolling Bearing with Cage Flexibility under Accelerating Conditions. *Mech. Syst. Signal Process.* **2021**, *150*, 107257. [CrossRef]
27. Park, S.; Kim, W.; Kim, S.I. A Numerical Prediction Model for Vibration and Noise of Axial Flux Motors. *IEEE Trans. Ind. Electron.* **2014**, *61*, 5757–5762. [CrossRef]
28. Qi, Y.; Dai, H. Influence of Motor Harmonic Torque on Wheel Wear in High-Speed Trains. *Proc. Inst. Mech. Eng. Part F J. Rail Rapid Transit* **2019**, *234*, 095440971983080. [CrossRef]
29. Guasch-Pesquer, L.; Youb, L.; González-Molina, F.; Zeppa-Durigutti, E. Effects of Voltage Unbalance on Torque and Current of the Induction Motors. In Proceedings of the IEEE 2012 13th International Conference on Optimization of Electrical and Electronic Equipment (OPTIM), Brasov, Romania, 24–26 May 2012.
30. Pustovetov, M.Y.; Pustovetov, K.M. The Electromagnetic Torque Ripple of Three-Phase Induction Electric Machine, Operating as a Part of the Auxiliary Electric Drive Onboard of A.C. Electric Locomotive—A Factor Contributing to the Failure of Bearings. In Proceedings of the IEEE 2018 International Multi-Conference on Industrial Engineering and Modern Technologies (FarEastCon), Vladivostok, Russia, 3–4 October 2018.
31. Wang, Z.; Wang, R.; Crosbee, D.; Allen, P.; Zhang, W. Wheel Wear Analysis of Motor and Unpowered Car of a High-Speed Train. *Wear* **2019**, *444–445*, 203136. [CrossRef]
32. Wang, Z.W.; Mei, G.M.; Xiong, Q.; Yin, Z.H.; Zhang, W.H. Motor Car-Track Spatial Coupled Dynamics Model of a High-Speed Train with Traction Transmission Systems. *Mech. Mach. Theory* **2019**, *137*, 386–403. [CrossRef]
33. Wu, P.B.; Guo, J.Y.; Wu, H.; Wei, J. Influence of DC-Link Voltage Pulsation of Transmission Systems on Mechanical Structure Vibration and Fatigue in High-Speed Trains. *Eng. Fail. Anal.* **2021**, *130*, 105772. [CrossRef]
34. Zhu, H.Y.; Yin, B.C.; Hu, H.T.; Xiao, Q. Effects of Harmonic Torque on Vibration Characteristics of Gear Box Housing and Traction Motor of High-Speed Train. *J. Traffic Transp. Eng.* **2019**, *19*, 65–76. (In Chinese)
35. Li, G.L. Research on Harmonic Suppression of Urban Rail Transit Traction Motor Based on Proportional Resonant Regulator. *Railw. Locomot. Car* **2019**, *39*, 111–115. (In Chinese)

36. Li, G.Q.; Liu, Z.M.; Guo, R.B. Stress Response and Fatigue Damage Assessment of High-Speed Train Gearbox. *J. Traffic Transp. Eng.* **2018**, *18*, 79–88. (In Chinese)
37. Liang, X.R.; Xiao, L.; Wang, X.Q. Design of Neural Network PID Controller for Speed Tracking of High-Speed Train. *Comput. Eng. Appl.* **2021**, *57*, 252–258. (In Chinese)
38. Xu, C.F.; Chen, X.Y.; Zheng, X. Slip Velocity Tracking Control of High-speed Train Using Dynamic Surface Method. *J. Railw.* **2020**, *42*, 41–49. (In Chinese)
39. Hou, T.; Guo, Y.Y.; Chen, Y. Study on Speed Control of High-Speed Train Based Multi-Point Model. *J. Railw. Sci. Eng.* **2020**, *17*, 314–325. (In Chinese)
40. Takahashi, I.; Noguchi, T. A New Quick-Response and High-Efficiency Control Strategy of an Induction Motor. *IEEE Trans. Ind. Appl.* **1986**, *22*, 820–827. [CrossRef]
41. Zhang, Z.; Tang, R.; Bai, B.; Xie, D. Novel Direct Torque Control Based on Space Vector Modulation with Adaptive Stator Flux Observer for Induction Motors. *IEEE Trans. Magn.* **2010**, *46*, 3133–3136. [CrossRef]
42. Habetler, T.G.; Profumo, F.; Pastorelli, M. Direct Torque Control of Induction Machines Using Space Vector Modulation. *Ind. Appl. IEEE Trans.* **1992**, *28*, 1045–1053. [CrossRef]
43. Jin, S.; Li, M.; Zhu, L.; Liu, G.; Zhang, Y. Direct Torque Control of Open Winding Brushless Doubly-Fed Machine. In Proceedings of the 2017 IEEE International Magnetics Conference (INTERMAG), Dublin, Ireland, 24–28 April 2017.
44. Casadei, D.; Serra, G.; Tani, K. Implementation of a Direct Control Algorithm for Induction Motors Based on Discrete Space Vector Modulation. *Power Electron. IEEE Trans.* **2000**, *15*, 769–777. [CrossRef]
45. Beerten, J.; Verveckken, J.; Driesen, J. Predictive Direct Torque Control for Flux and Torque Ripple Reduction. *IEEE Trans. Ind. Electron.* **2010**, *57*, 404–412. [CrossRef]
46. Stojic, D.; Milinkovic, M.; Veinovic, S.; Klasnic, I. Improved Stator Flux Estimator for Speed Sensorless Induction Motor Drives. *Power Electron. IEEE Trans.* **2014**, *30*, 2363–2371. [CrossRef]
47. Buja, G.S.; Kazmierkowski, M.P. Direct Torque Control of PWM Inverter-Fed AC Motors—A Survey. *IEEE Trans. Ind. Electron.* **2004**, *51*, 744–757. [CrossRef]
48. Demir, R.; Barut, M.; Yildiz, R.; Inan, R.; Zerdali, E. EKF Based Rotor and Stator Resistance Estimations for Direct Torque Control of Induction Motors. In Proceedings of the 2017 International Conference on Optimization of Electrical and Electronic Equipment (OPTIM) & 2017 Intl Aegean Conference on Electrical Machines and Power Electronics (ACEMP), Brasov, Romania, 25–27 May 2017.
49. Barnes, M. Practical Variable Speed Drives and Power Electronics. *Pract. Variab. Speed Drives Power Electron.* **2003**, *4*, 156–177.
50. Faiz, J.; Hossieni, S.H.; Ghaneei, M. Direct Torque Control of Induction Motors for Electric Propulsion System. *Electr. Power Syst. Res.* **1999**, *51*, 95–101. [CrossRef]
51. Vaez-Zadeh, S.; Jalali, E. Combined Vector Control and Direct Torque Control Method for High Performance Induction Motor Drives. *Energy Convers. Manag.* **2007**, *48*, 3095–3101. [CrossRef]
52. Baader, U.; Depenbrock, M.; Gierse, G. Direct Self Control of Inverter-Fed Induction Machine, a Basis for Speed Control without Speed-Measurement. In Proceedings of the Conference Record of the IEEE Industry Applications Society Annual Meeting, San Diego, CA, USA, 1–5 October 1989.
53. Rao, Z. *Calculation of Train Traction*; China Railway Press: Beijing, China, 1997. (In Chinese)
54. Wen, X.Y. *Research on Traction Motor Optimization Control for Low Speed Operation of Locomotive*; Beijing Jiaotong University: Beijing, China, 2012. (In Chinese)

Article

Mechanical Wear Contact between the Wheel and Rail on a Turnout with Variable Stiffness

Jerzy Kisilowski [1] and Rafał Kowalik [2],*

[1]   Faculty of Transport, Electrical Engineering and Computer Science, University of Technology and Humanities, 26-600 Radom, Poland; jerzy@kisilowscy.waw.pl

[2]   Department of Avionics and Control Systems, Faculty of Aviation Division, Military University of Aviation, 08-521 Deblin, Poland

*   Correspondence: r.kowalik@law.mil.pl; Tel.: +48-26-151-88-24

Abstract: The operation and maintenance of railroad turnouts for rail vehicle traffic moving at speeds from 200 km/h to 350 km/h significantly differ from the processes of track operation without turnouts, curves, and crossings. Intensive wear of the railroad turnout components (switch blade, retaining rods, rails, and cross-brace) occurs. The movement of a rail vehicle on a switch causes high-dynamic impact, including vertical, normal, and lateral forces. This causes intensive rail and wheel wear. This paper presents the wear of rails and of the needle in a railroad turnout on a straight track. Geometrical irregularities of the track and the generation of vertical and normal forces occurring at the point of contact of the wheel with turnout elements are additionally considered in this study. To analyse the causes of rail wear in turnouts, selected technical–operational parameters were assumed, such as the type of rail vehicle, the type of turnout, and the maximum allowable axle load. The wear process of turnout elements (along its length) and wheel wear is presented. An important element, considering the occurrence of large vertical and normal forces affecting wear and tear, was the adoption of variable track stiffness along the switch. This stiffness is assumed according to the results of measurements on the real track. The wear processes were determined by using the work of Kalker and Chudzikiewicz as a basis. This paper presents results from simulation studies of wear and wear coefficients for different speeds. Wear results were compared with nominal rail and wheel shapes. Finally, conclusions from the tests are formulated.

Keywords: wear; turnout; rail; stiffness; high-speed

Citation: Kisilowski, J.; Kowalik, R. Mechanical Wear Contact between the Wheel and Rail on a Turnout with Variable Stiffness. *Energies* **2021**, *14*, 7520. https://doi.org/10.3390/en14227520

Academic Editor: Tek Tjing Lie

Received: 3 October 2021
Accepted: 8 November 2021
Published: 11 November 2021

Publisher's Note: MDPI stays neutral with regard to jurisdictional claims in published maps and institutional affiliations.

Copyright: © 2021 by the authors. Licensee MDPI, Basel, Switzerland. This article is an open access article distributed under the terms and conditions of the Creative Commons Attribution (CC BY) license (https://creativecommons.org/licenses/by/4.0/).

1. Introduction

The study of wear in wheel-track systems is the subject of many works [1–11]. In these works, the task of wear is addressed by presenting different models of wheel–rail interaction. In most works, a constant value of normal force was assumed. In some works, simulations of rail vehicle movement on a straight track without a turnout were performed. Motion without a turnout and motion through a turnout were considered with track susceptibility (elasticity and viscous damping) as constant. According to the research conducted in [12], there are significant differences in the appearing vertical forces and normal forces when passing through a turnout with different values of the beam–subtrack system susceptibility. These forces affect the process of phenomena in wheel–rail contact and have an impact on the wear in the wheel–rail pair. The wear phenomenon was studied based on the works [13,14]. Simulation of rail vehicle movement on a straight track without a turnout and a track with a turnout was also shown.

The infrastructure of a rail transportation system consists of railroad tracks, curves, intersections, and turnouts. Turnouts are a complex structure of railroads. They connect neighbouring tracks and enable railway vehicles to change direction of travel. The basic type is the ordinary turnout consisting of switch blades (2), closure rails (2, 3), a crossing frog (4), turnout sleepers, and setting devices. The crucial element of each turnout is the frog (4) that enables the wheels of railway vehicles to roll over the place of rail crossing. Due to difficult operating conditions characterised by high-dynamic loads generated by the wheels of rail vehicles, the crossbeams are particularly exposed to the destructive character of impact loads [15–20]. Crossings can have a fixed or movable bow. The subject of the analysis is a right-hand ordinary turnout with a radius of R = 1200 m, a fixed crossbeam of 1:9, and three setting closures with a holding force of 7.5 kN (each). The individual elements of this turnout are shown in Figure 1.

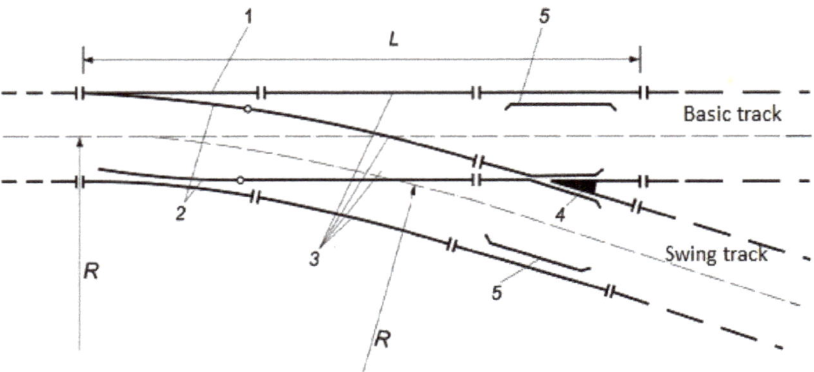

Figure 1. Normal right turnout 1—Stock rail, 2—Switch blades, 3—Closure rails, 4—Frog, 5—Guardrail.

The mechanical destruction process of the surface layer leads to undesirable changes in the dimensions and shape of the contacting rolling surfaces of the turnout and railroad wheel elements. Degradation of turnouts, especially crossbucks, contributes to the increase in dynamic interactions, which has an adverse effect on the cooperation of the wheel–rail system [21–26].

Railroad turnouts are particularly exposed to abrasive and fatigue wear, which causes shape changes that result from the impact of high-dynamic loads of cyclic nature that occur during the passage of rail vehicles [27,28].

Railway turnouts are important elements of railway infrastructure that ensure traffic runs smoothly between different branch tracks [29]. A turnout is a structure that allows railway vehicles to pass from one track to another while maintaining a certain speed [30]. One of the most common railway turnouts is the regular turnout (Rz) [31]. It consists of three basic units, such as the switch train assembly, the connecting rail assembly, and the crossover assembly. The switch assembly is a movable turnout unit that moves the switch blades by means of a drive. A smooth and safe track change depends on the correct execution of the initial part of the switch blade, which must have the appropriate shape in order to adequately adhere to the supporting rail in the switch. In turnouts, there are often two wheel–rail contact surfaces, as well as disturbances in the nominal wheel–rail contact conditions due to wheel movement from the main rail to the switch rail [32]. The dynamic interaction between a rail vehicle and a railway turnout is more complex than that of normal or curved tracks. Severe shock loads may occur during the passage through the turnout, generating severe damage to the surfaces of the turnout components [33]. Traffic of rail vehicles in regular operation may be considered as a source of influence of high-dynamic loads of cyclic nature, which translates into damages in the form of abrasive and fatigue wear and tear, as well as changes in the shape and dimensions of the outer layer [34–38]. Calculations of dynamic loads and resulting contact and internal stresses enable rational design of railway turnouts and correct selection of material to construct their elements [39]. The results described in [40] show that profile wear disturbs the distribution of wheel–rail contact points, changes the position of wheel–rail contact points along the longitudinal direction, and affects the dynamic interaction between the vehicle and the turnout. Profile wear disturbs the normal contact situations between the wheel and switch rail and worsens the stress condition of the switch rail [41]. This model allowed the rational design of railway turnouts and the correct selection of material from which their components are made [42].

The development of turnout constructions also results from technological progress in the production of new steel grades for railway turnouts, the development of new material testing methods, and a better understanding of the phenomena occurring in wheel–rail interaction [43–48].

2. Mathematical Model of the Rail Vehicle–Track System

To determine the durability of individual elements of the railroad switch, it is necessary to calculate the characteristics of the load in the function of time and distance, originating from the wheelsets of the railroad vehicles acting along the switch. The method used to determine the distribution of forces along the switch is the simulation of mathematical models showing the dynamics of the rail vehicle–track system using computer software. When modelling the dynamics of the wheel–rail system, the most used programs are MATLAB and Universal Mechanism. The Universal Mechanism program provides greater capabilities in modelling dynamic phenomena.

In the modelling process, a high-speed train was used, the parameters of which were taken from the work [19].

The mathematical model of the rail vehicle was built based on linear and angular coordinate systems shown in Figure 2.

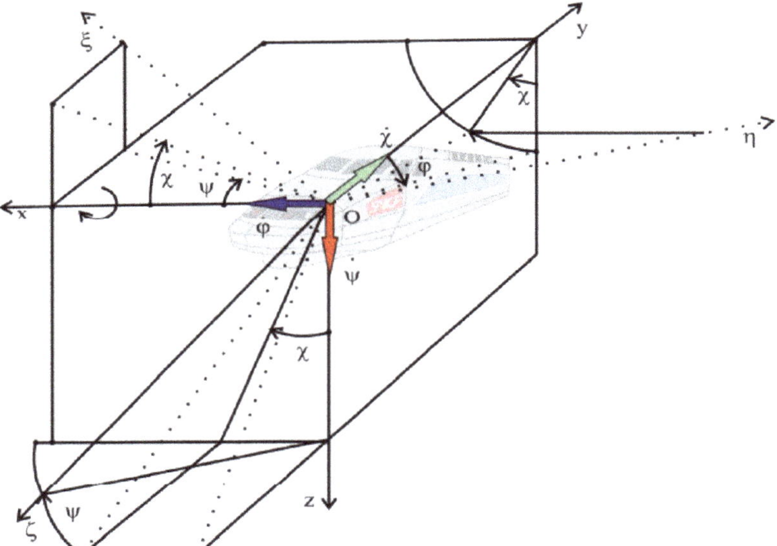

Figure 2. Linear and angular coordinate system.

This system is used as an inertial system associated with rigid solids, of which three groups can be listed in a vehicle (typical) as shown in Figure 3.

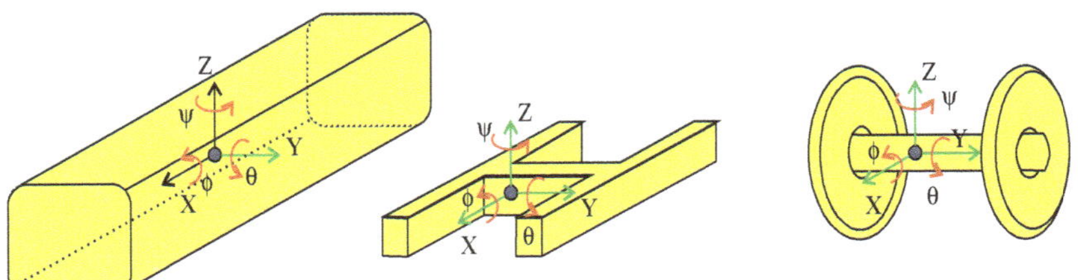

Figure 3. Coordinate systems in rigid bodies of a rail vehicle.

The vehicle has a body, two bogies, and four sets of wheels. The coordinate systems originate at the centre of mass of the individual solids, and the axes lie on the axes of symmetry. There is an identical system associated with the track that is called a non-inertial system. The matrix of directional cosines between the inertial and non-inertial systems was assumed to be zero–one [14,15].

In addition, the equations of the ties were assumed according to the coordinate system and Figure 3.

The ties for analysing the system can be written as follows (Figure 4):

$$\Phi = \frac{z_p - z_l}{2b}, z = \frac{z_p - z_l}{2},$$

$$z_{tl} = z_l - z_{wl} - \left(y - \frac{y_{wl}}{2}\right)\sigma, \qquad z_{tp} = z_p - z_{wp} + \left(y - \frac{y_{wp}}{2}\right)\sigma, \tag{1}$$

$$\dot{\chi} = -\frac{1}{r}\dot{x}$$

where $2b$ is the distance between the contact points (wheel and rail) in the middle position of the wheelset; r is the radius of the wheel included in the wheelset, measured in the middle position; σ is the coefficient linking the angular and transverse displacement of a wheelset; and z_{tp}, z_p, z_{wp}, z_{tl}, z_l, and z_{wl} are auxiliary nodes, used in the mathematical description of the movement of the railway vehicle.

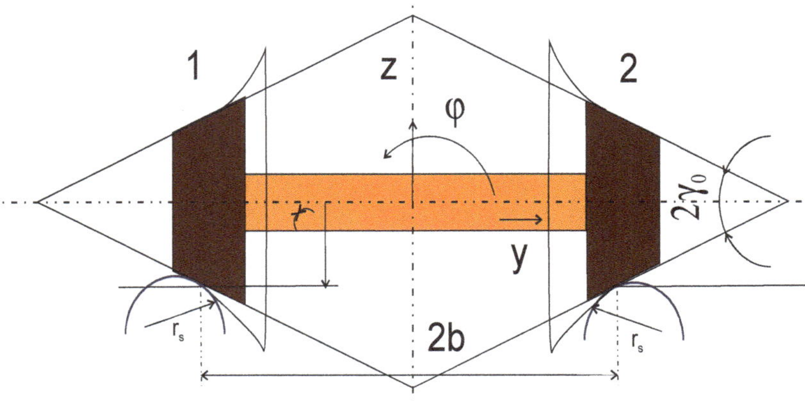

Figure 4. Geometry model of the wheelset–track system (wheelset in middle position) where 1—Outer rail (left), 2—Inner rail (right) [16].

The following assumptions were made in developing the equations of motion:

The vertical loads occurring on the rail will be a variable value and will be determined from the previous step of mathematical calculations performed to determine the train parameters (wheelset and bogie spacing).

The railroad track was modelled as a Euler–Bernoulli beam on which a wheel with velocity v is rolling (motion on the straight track and motion on the turnout return track were considered).

The contact between the rolling surfaces of the wheels and the rail heads is defined based on Kalker's linear theory (defining ellipses with semi axes *a* and *b*).

In the wheel–rail contact area, the Coulomb kinetic sliding friction with a constant coefficient of friction is considered.

Such phenomena as adhesion, micro-slip, and material wear of the wheel and the rail were also considered in the dynamics of vehicle movement on the track.

In the model under consideration, the possibility of two contact ellipses occurring because of two-point rolling of the wheel on the rail within the turnout has been taken into account.

Suspension elements of the first and second stage were assumed to be linear for all assumed coordinates.

For the track without a turnout, the susceptibility was assumed according to the Voigt model (linear stiffness and linear damping). These quantities were determined by measurements on the actual object and consist of the stiffness coefficient, 0.2×10^8 N/m, and damping coefficient, 3.2×10^3 Ns/m [16].

At the turnout, the stiffness coefficient was calculated according to the parameters shown in Figure 5.

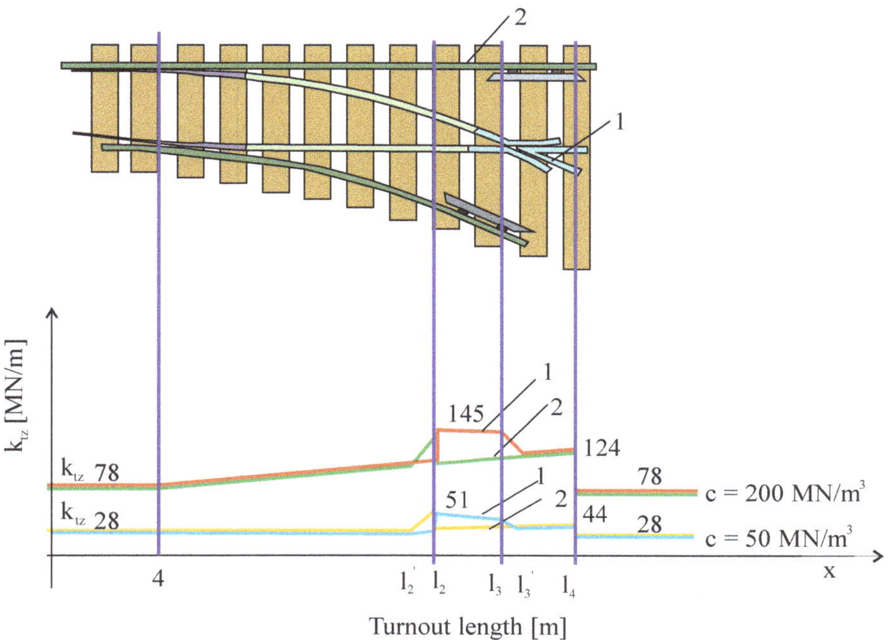

Figure 5. The course of variation of the vertical stiffness coefficient of the rails in the switch with different values of the bedding coefficient (real measurements on CMK Idzikowice): 1—Inner track (with cross-brace), 2—Outer track [18].

Considering geometric and structural constraints, the rail vehicle has 27 degrees of freedom. The equation can be found in the work [19].

3. Results

The aim of this paper is to analyse the wear of a railway turnout with a radius of 3000 m, considering the change in contact area resulting from the variation of normal force. For the guidance mechanism of a wheelset on a through track, if the wheelset is offset to the side, the wheel radii at the wheel contact point are different. Due to the rigid coupling of speeds, one wheel becomes the driving wheel, and the other wheel becomes the braking wheel. This leads to a "steering motion" which drives the wheelset back to the centre of the track. The movement continues past the track axis until the same situation occurs in mirror image to the starting position; then the process starts again. It should be noted that during the passage through the crossing area, there are sleepers laid next to each other that are connected with a change in the substrate stiffness. The next stage of analysis on rail vehicle motion is the passage on a diverging track where, despite the rigid speed coupling between the wheels rolling on the inside and outside of the curve, the wheelset can turn without slipping on curves with large radii. This is possible because the lateral displacement towards the outer rail of the curve turnout results in a difference in wheel radii, which means that the peripheral velocity at the point of contact for the outer wheel is greater than that of the inner wheel. Bearing in mind the discussed phenomena, an attempt has been made to investigate the change in the value of normal force for wear that occurs in railway turnouts.

Using Universal Mechanism and MATLAB software, simulations were performed to determine vertical forces and normal forces for speeds from 100 km/h to 350 km/h, shown in Figures 6–17.

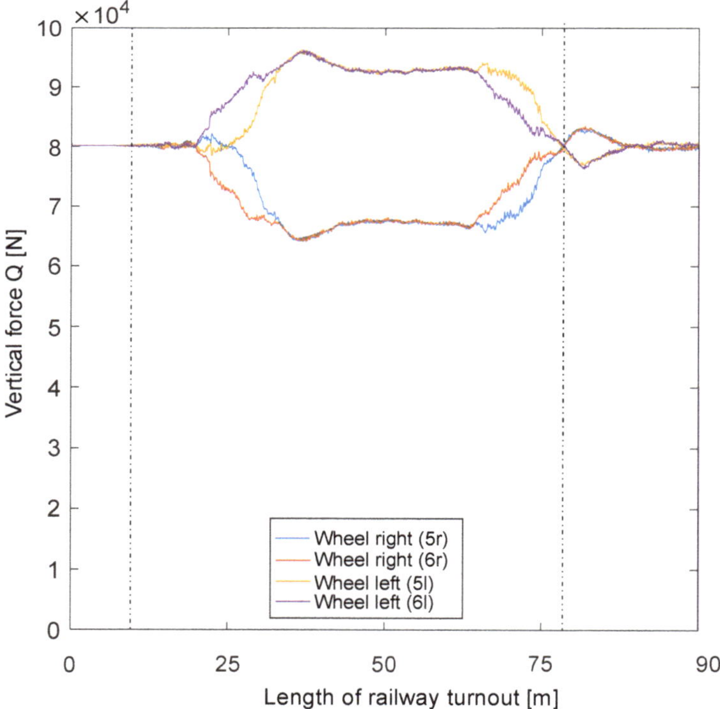

Figure 6. Vertical force on wheels at 100 km/h on straight track through a turnout.

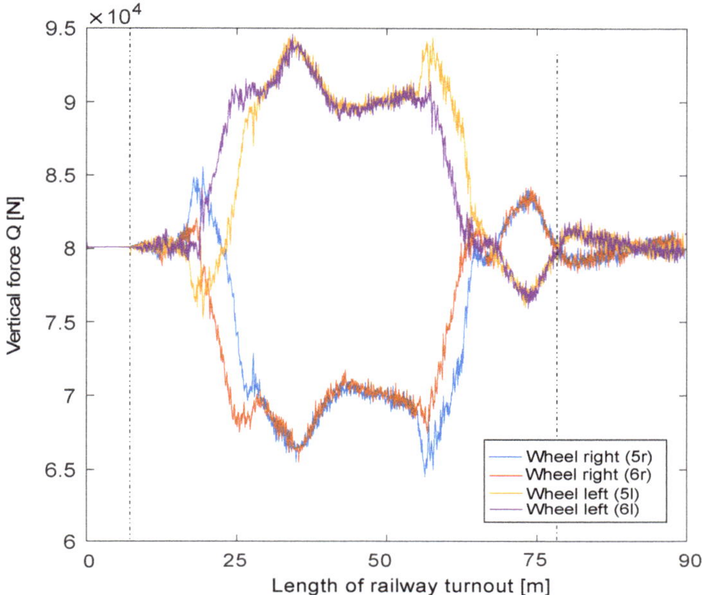

Figure 7. Vertical force on wheels at 150 km/h on a straight track through a turnout.

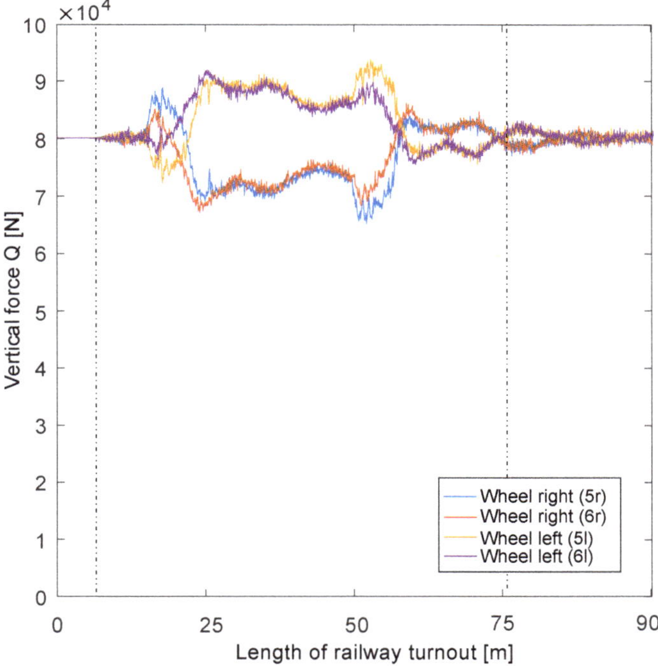

Figure 8. Vertical force on wheels at 200 km/h on a straight track through a turnout.

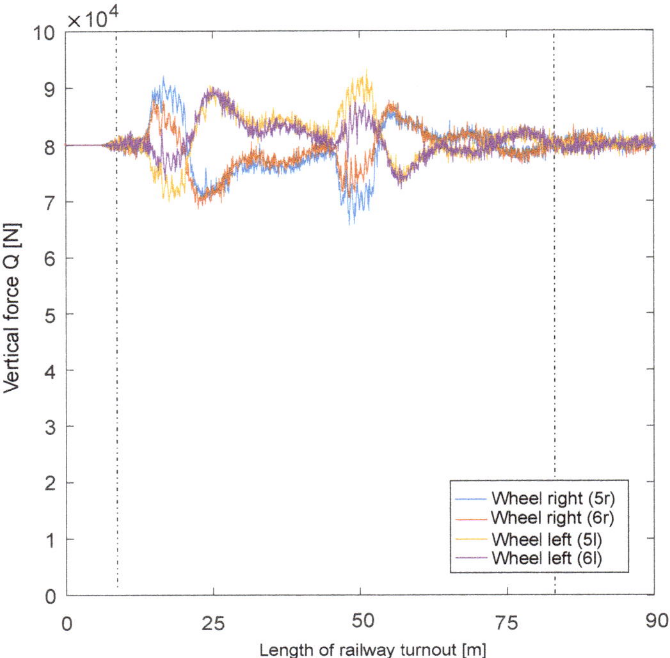

Figure 9. Vertical force on wheels at 250 km/h on a straight track through a turnout.

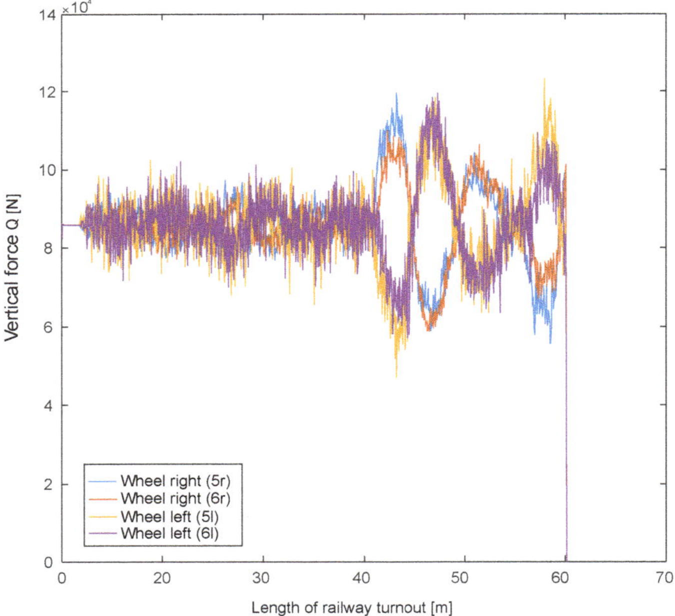

Figure 10. Vertical force at 300 km/h on a straight track through a turnout.

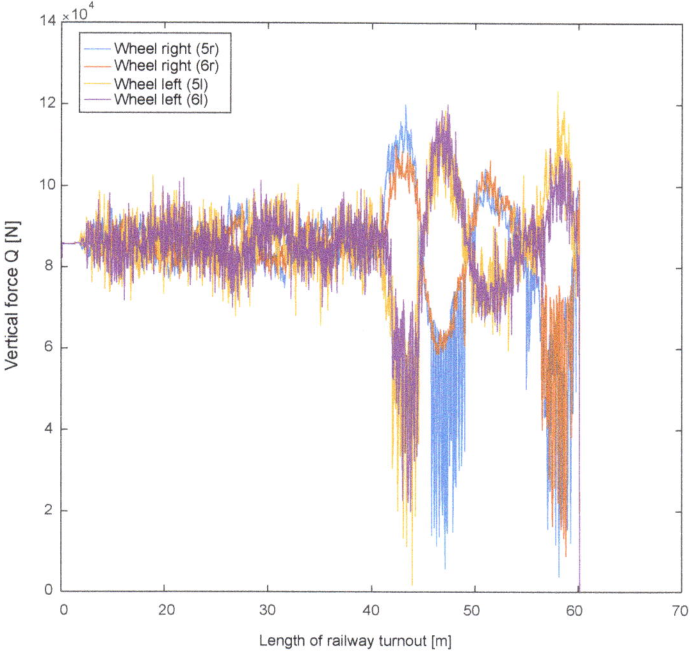

Figure 11. Vertical force at 350 km/h on a straight track through a turnout.

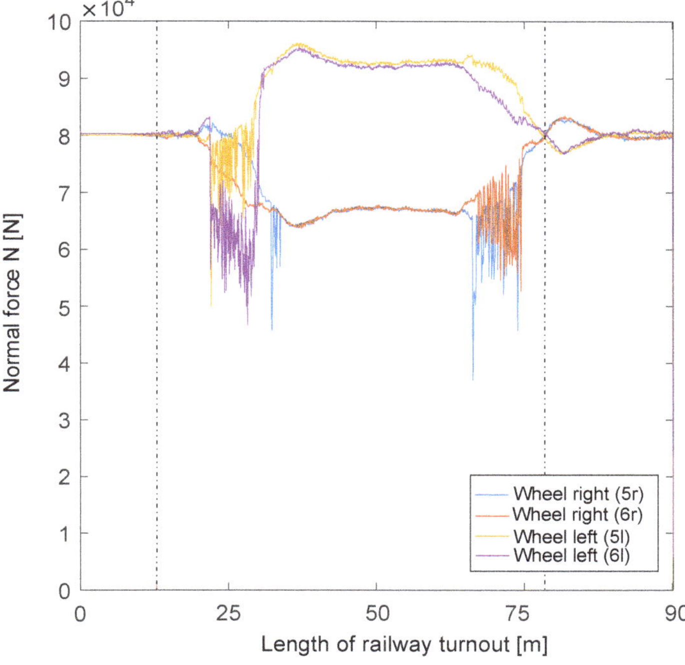

Figure 12. Normal force at 100 km/h on a straight track through a turnout.

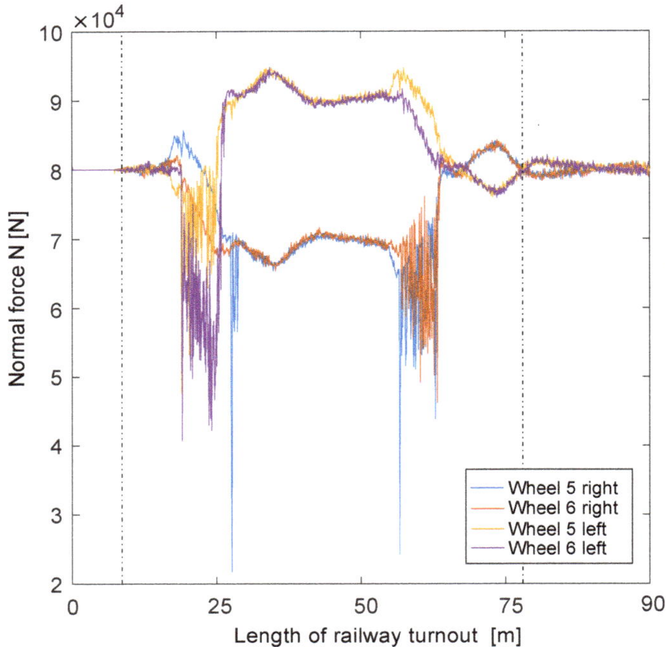

Figure 13. Normal force at 150 km/h on a straight track through a turnout.

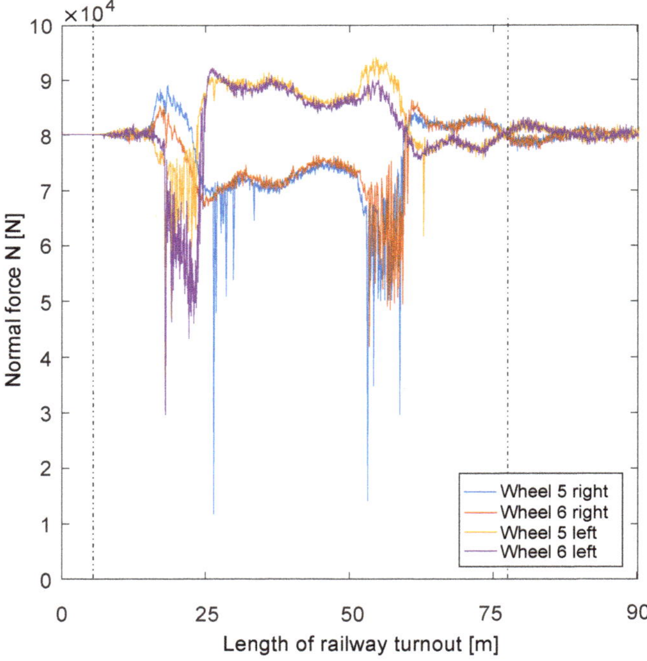

Figure 14. Normal force at 200 km/h on a straight track through a turnout.

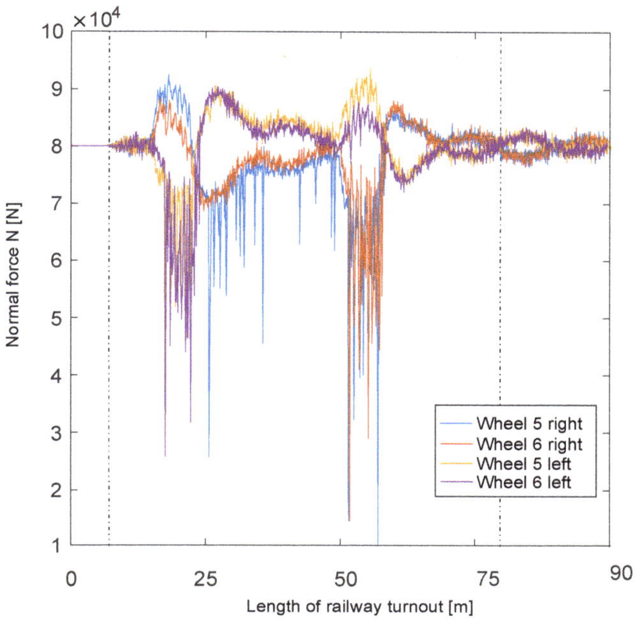

Figure 15. Normal force at 250 km/h on a straight track through a turnout.

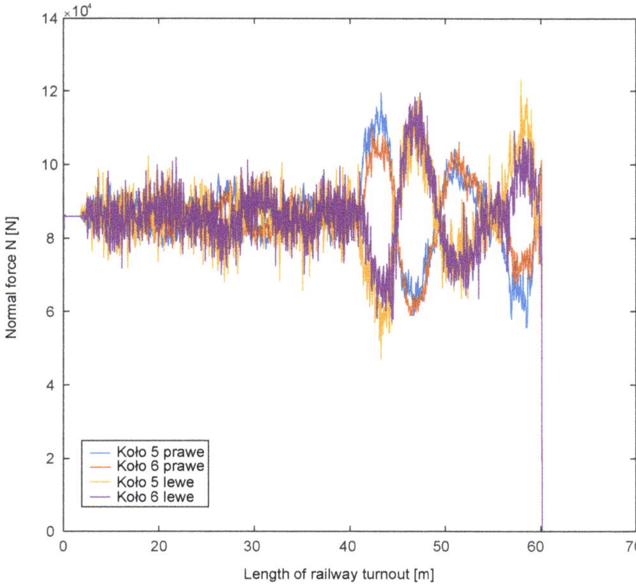

Figure 16. Normal force at 300 km/h on a straight track through a turnout.

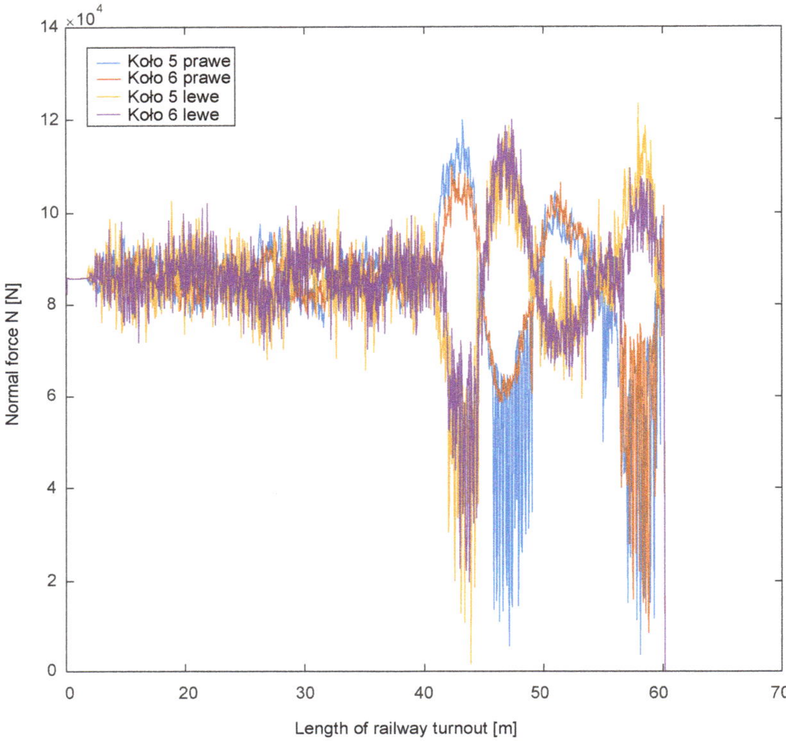

Figure 17. Normal force at 350 km/h on a straight track through a turnout.

As seen in the figures, the magnitudes of vertical and normal forces increase as velocity increases. Significant changes occur if speeds increase above 100 km/h. For speeds above 250 km/h, the increase in vertical and normal forces occurs mainly within the cross member. The change in these quantities is due to the change in track stiffness. For a straight track without a turnout, the stiffness of the track is usually assumed constant. For a track with a turnout, the stiffness changes along the length as shown in Figure 4. These quantities increase up to 1.5 times the static load.

Next, simulations were performed to determine the vertical and normal forces in straight-track traffic without a turnout. This was done for speeds from 150 km/h to 350 km/h and is shown in Figures 18–27. The value of the force due to the load per wheel is 8.1×10^4 N. The alternating loads fluctuate around this value.

From the results presented, there are oscillations of these forces in the movement on the track without turnout, but there is no significant difference. There is an increase of about 1×10^4 N in the normal force. These forces are the basis for determining the contact surfaces and the amount of wear.

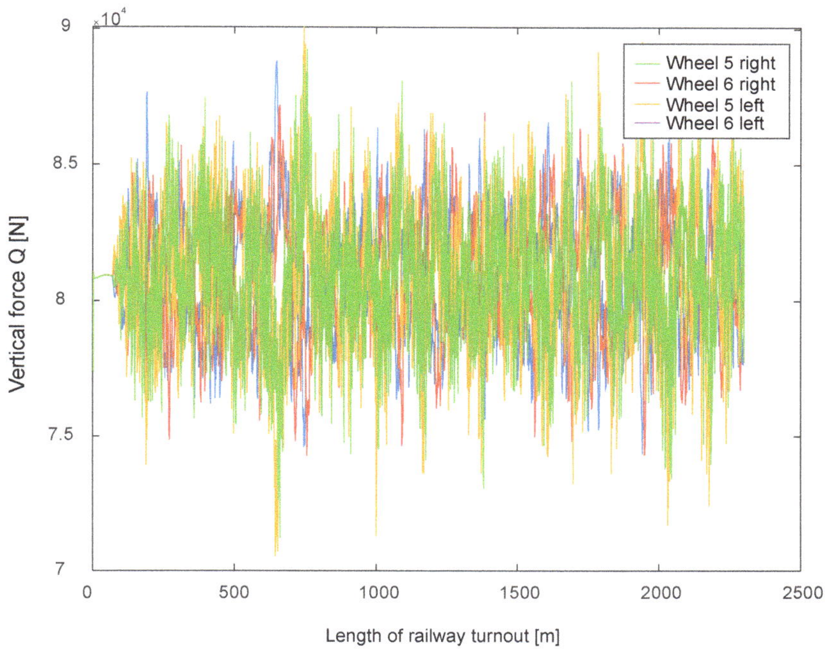

Figure 18. Vertical force at 150 km/h on a straight track without a turnout.

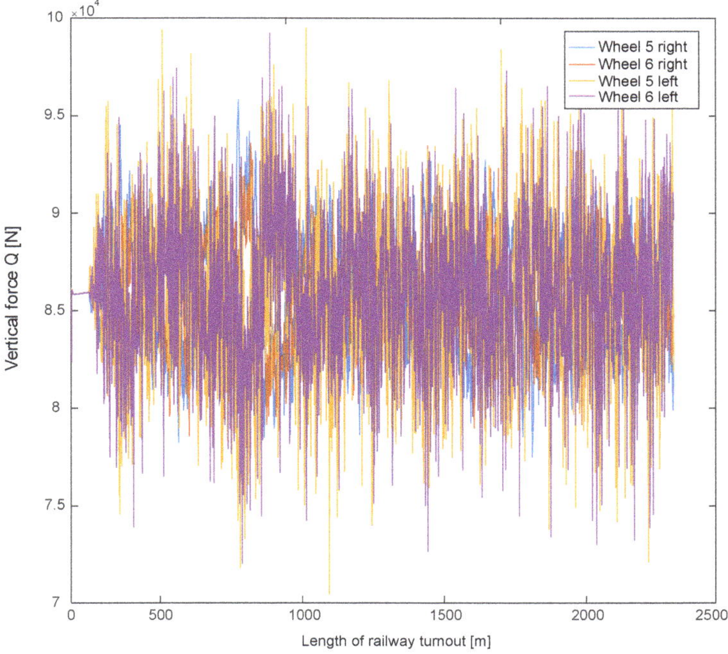

Figure 19. Vertical force at 200 km/h on a straight track without a turnout.

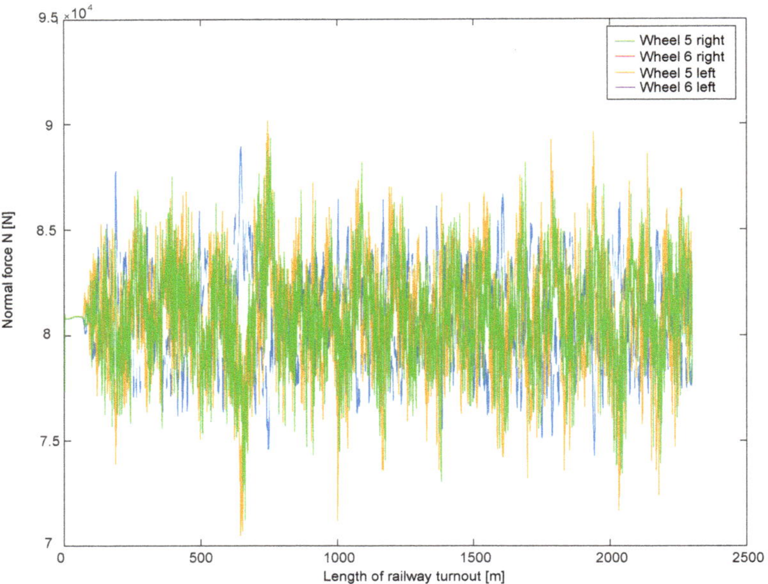

Figure 20. Normal force at 150 km/h on a straight track without a turnout.

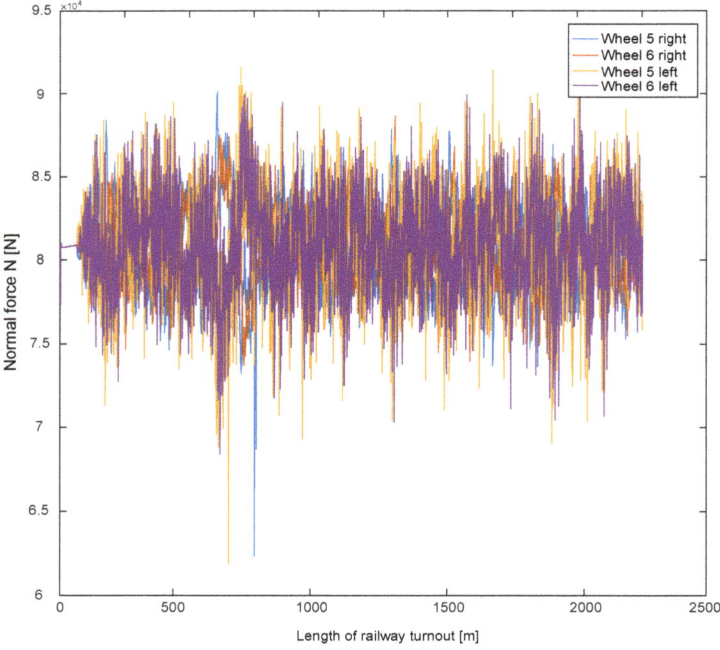

Figure 21. Normal force at 200 km/h on a straight track without a turnout.

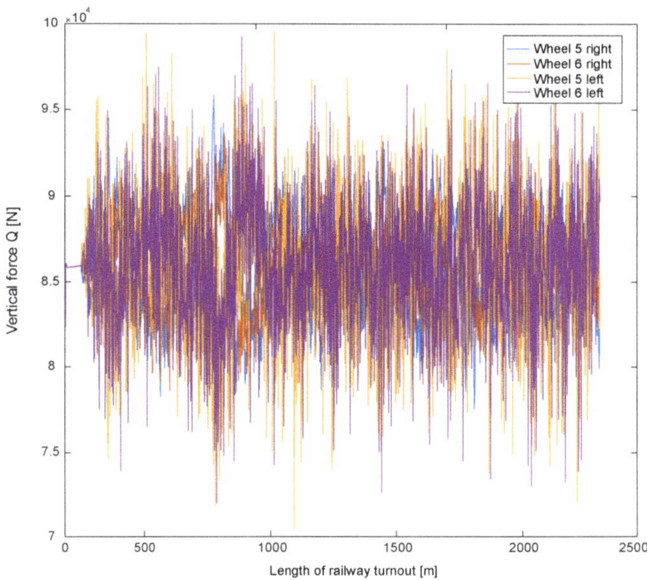

Figure 22. Vertical force at 250 km/h on a straight track without a turnout.

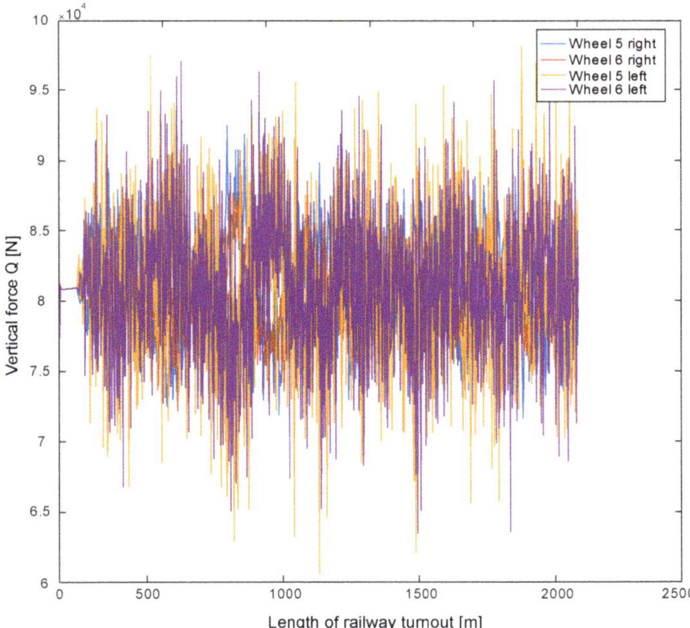

Figure 23. Vertical force at 300 km/h on a straight track without a turnout.

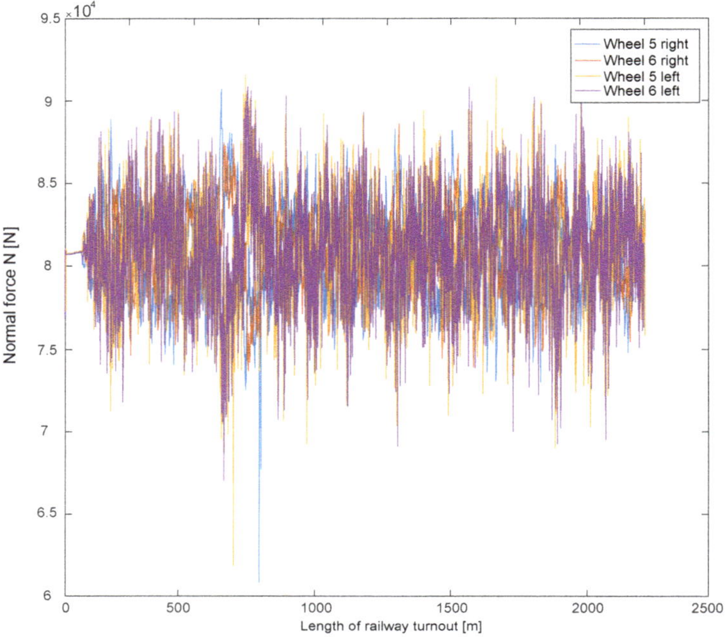

Figure 24. Normal force at 250 km/h on a straight track without a turnout.

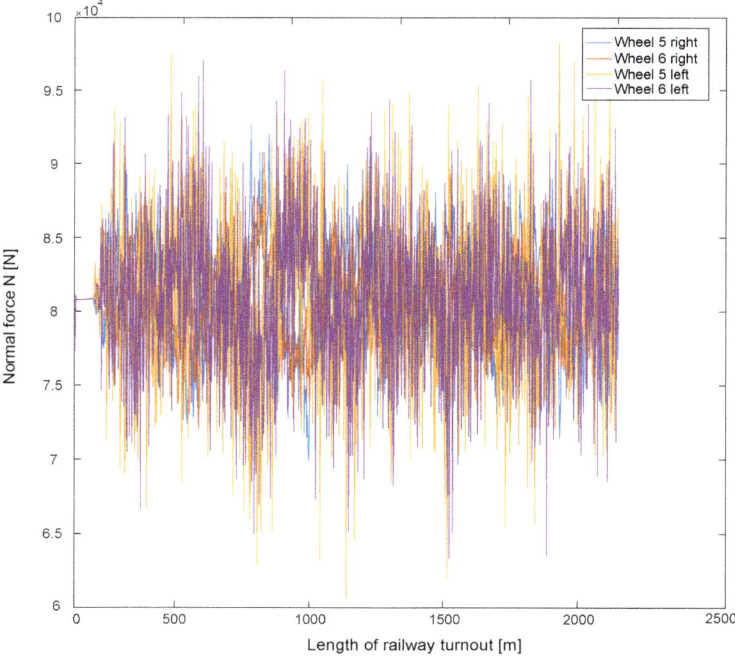

Figure 25. Normal force at 300 km/h on a straight track without a turnout.

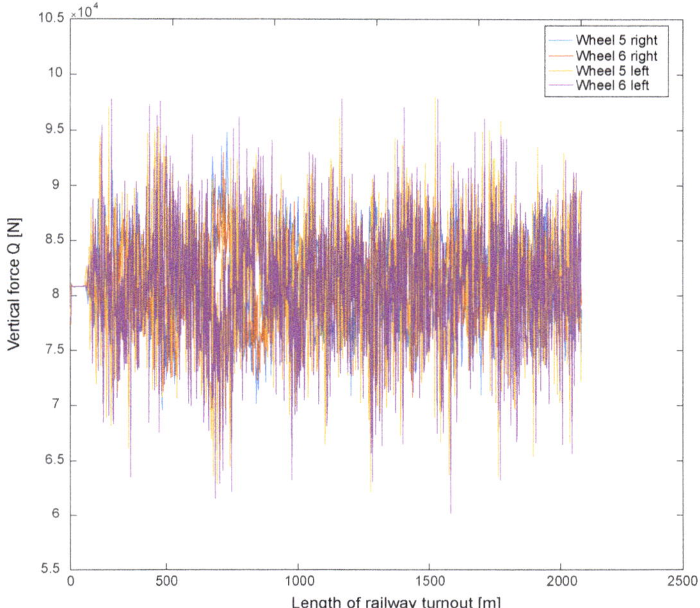

Figure 26. Vertical force at 350 km/h on a straight track without a turnout.

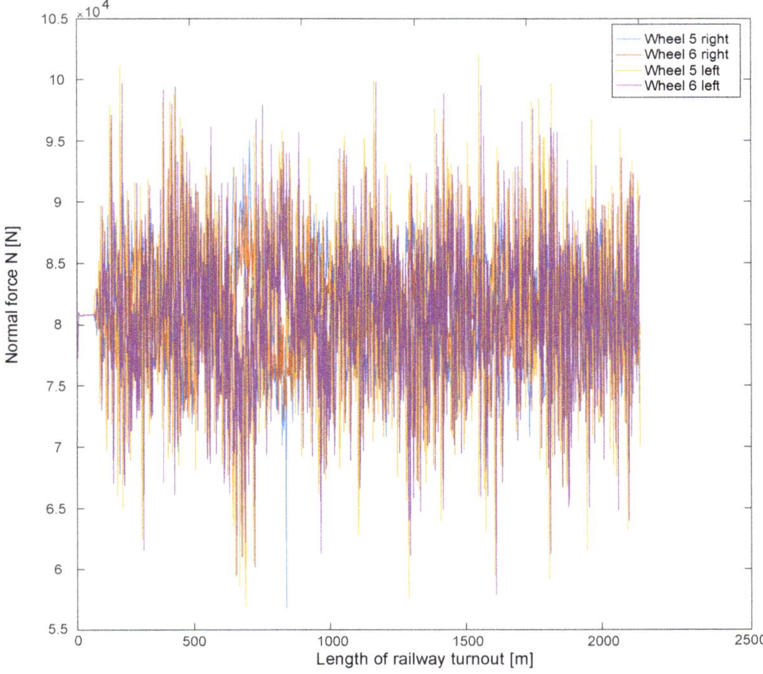

Figure 27. Normal force at 350 km/h on a straight track without a turnout.

Next, we proceeded to determine the wear on the wheel and rail. Two works, [12] and [13], were used to consider this topic. Based on them, the following relations can be written:

$$W_m = C \cdot W_f$$
$$W_d = C_1 \cdot W_{f-na}$$

(2)

where W_m is the mass consumed [$\mu \cdot$ g] per unit contact surface, W_d is the depth of the wear surface [mm], C is constant (for steel ≈ 0.00124 $\mu \cdot$ g/N $\cdot$ mm), W_f is the work of friction forces [N $\cdot$ mm], C_1 is the constant $\approx 1.55 \times 10^{-7}$ [mm/N], and W_{f-na} is the work of friction forces per unit contact surface [mm^2/Nn].

$$m_W = C \cdot W_l$$

(3)

where m_W is the mass of consumed material per unit contact area [$\mu \cdot$ g /mm^2], C is the constant (for steel ≈ 0.00124 $\mu \cdot$ g/N $\cdot$ mm), and W_l is the work done by friction force per unit area of contact ellipse [mm^2/N].

Using the presented relations, W_l was determined as the wear factor. The software used to perform the simulation makes it possible to determine the wear factor. Such a test was performed for a straight track with a turnout and for a straight track without a turnout. The test results are shown in Figures 28–37. The figures show wear factors and wheel and rail wear for the passage of a rail vehicle through a turnout with and without a turnout.

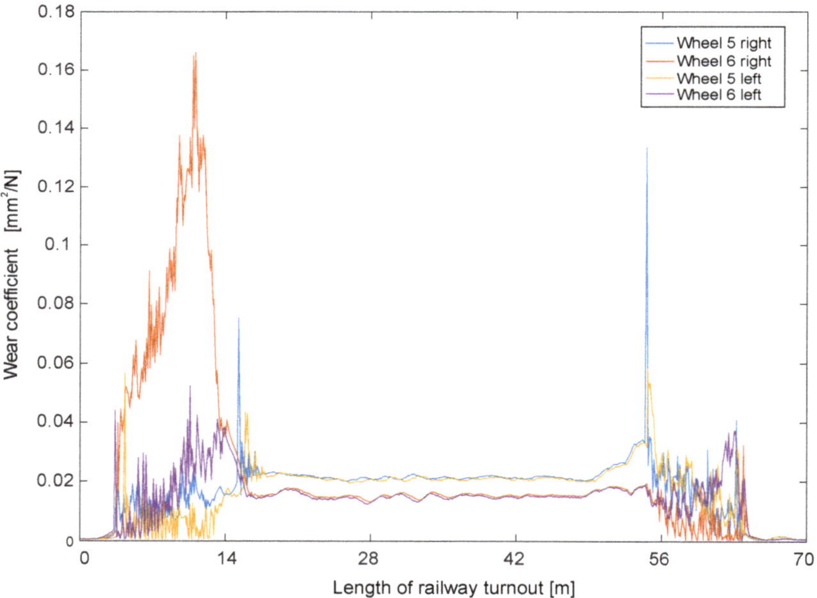

Figure 28. Wear coefficient at 100 km/h on a straight track with a turnout.

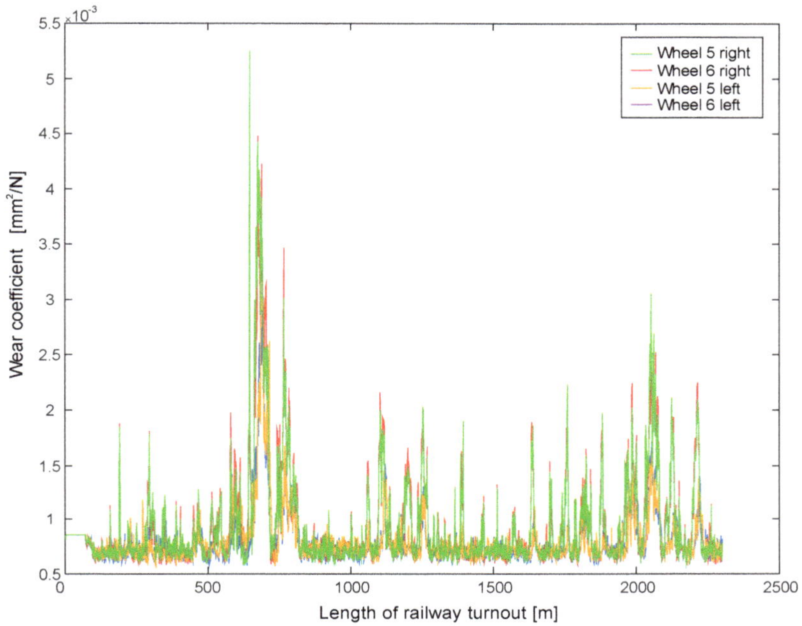

Figure 29. Wear coefficient obtained for a straight track without a turnout at 150 km/h.

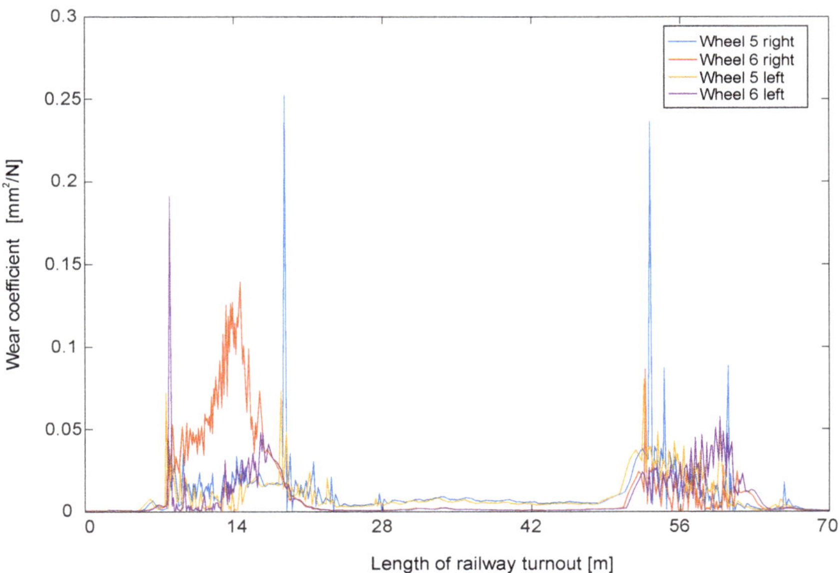

Figure 30. Wear coefficient at 200 km/h on a straight track with a turnout.

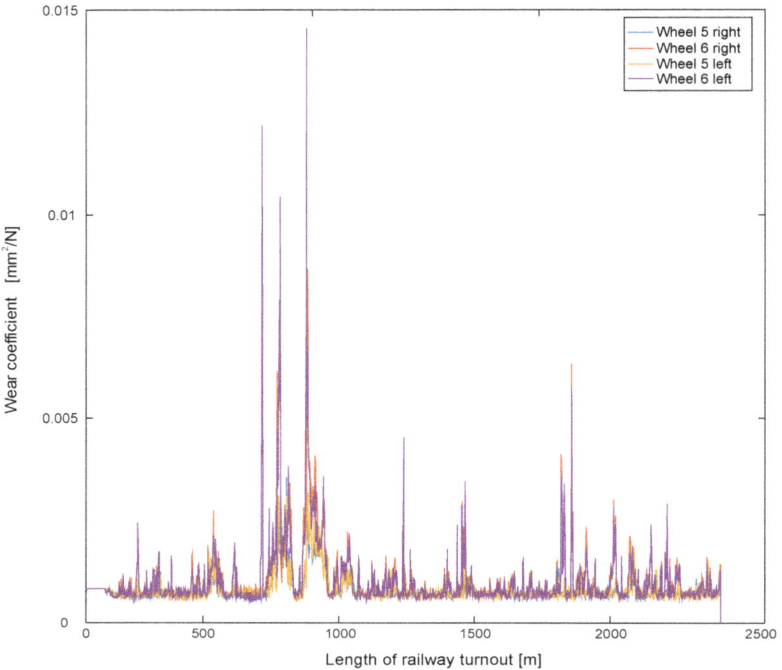

Figure 31. Wear coefficient obtained for a straight track without a turnout at 200 km/h.

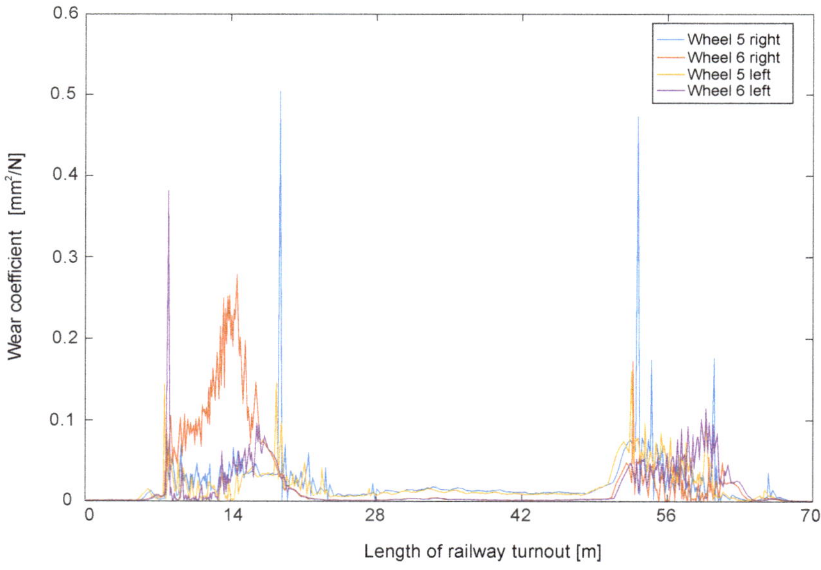

Figure 32. Wear coefficient obtained for a straight track with a turnout at 250 km/h.

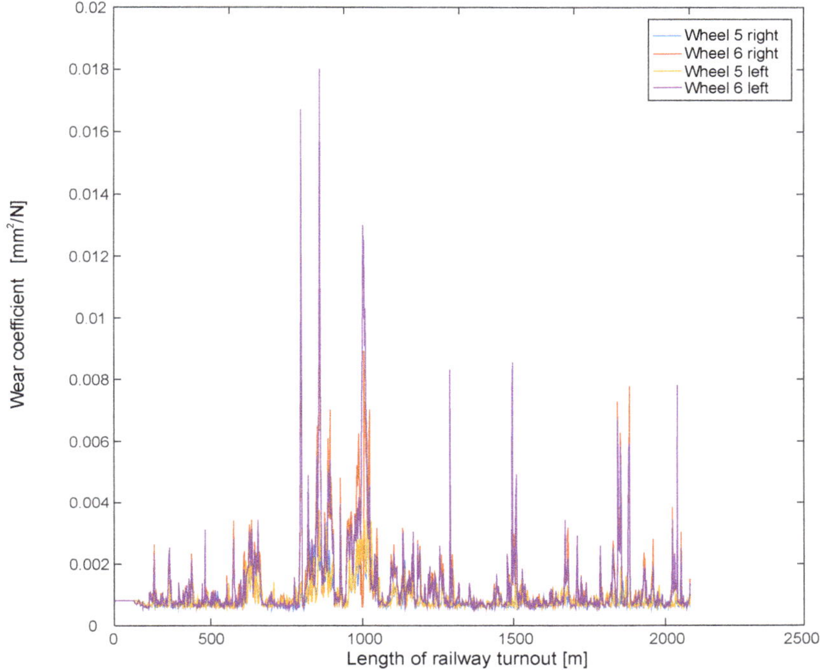

Figure 33. Wear coefficient obtained for a straight track without a turnout at 250 km/h.

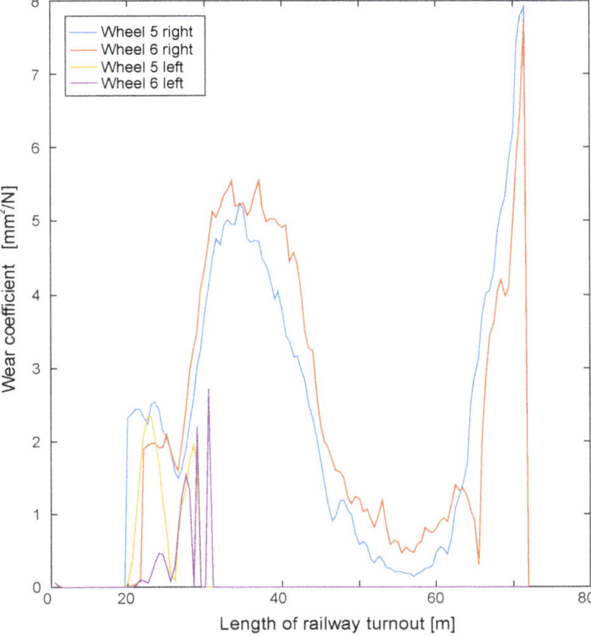

Figure 34. Wear coefficient obtained for a straight track at 300 km/h with a turnout.

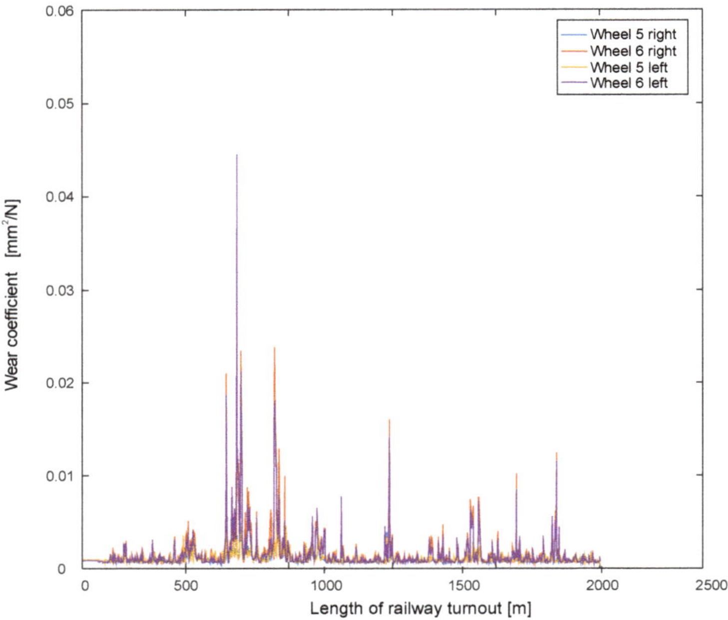

Figure 35. Wear coefficient obtained for a straight track without a turnout at 300 km/h.

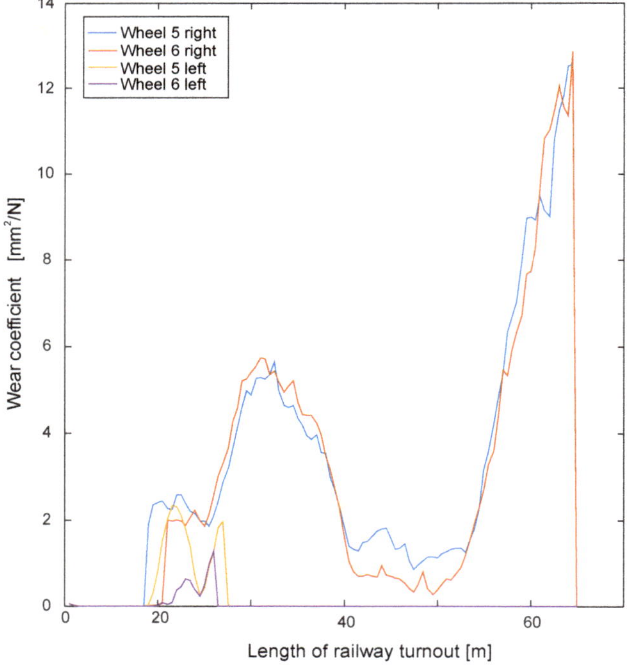

Figure 36. Wear coefficient obtained for a straight track at 350 km/h with a turnout.

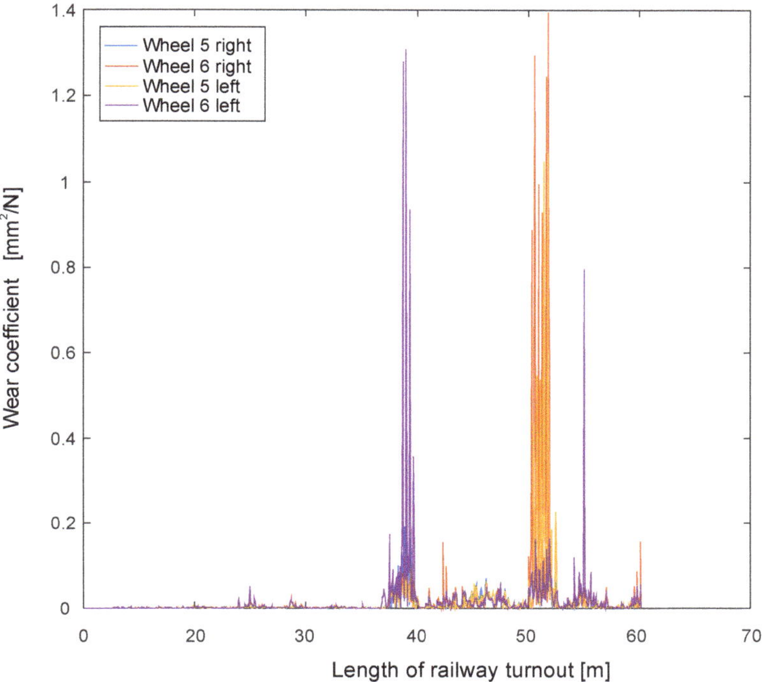

Figure 37. Wear coefficient obtained for a straight track without a turnout at 350 km/h.

From the simulations presented, when passing through a turnout, the wear coefficients increase in the turnout entry area (needle and resistor) and in the crossover area. For traffic on a track without a turnout, the wear factors vary between 0.0005 and 0.05 N/mm^2, while for traffic through a turnout, the factors vary between 0.16 and 12 N/mm^2. In traffic on the track without a turnout, the maximum magnitudes come from the normal forces, which increase above the nominal force and decrease when the normal force decreases below the nominal force. The nominal force is 8.1×10^4 N.

Simulations of 20,000 train runs on a straight track through a turnout at 200 km/h were performed. The wear results are shown for the left wheel of the wheelset and the wear of the rail by the left wheel. The wear of the left wheel is shown in Figure 38, and the wear of the rail is shown in Figure 39. The wears for the other wheels and rails are identical.

For the same speed of 200 km/h, simulations were performed for wheel and rail wear when the rail vehicle moves on the track without turnout. Wheel wear is shown in Figure 40 and rail wear is shown in Figure 41 for the same conditions.

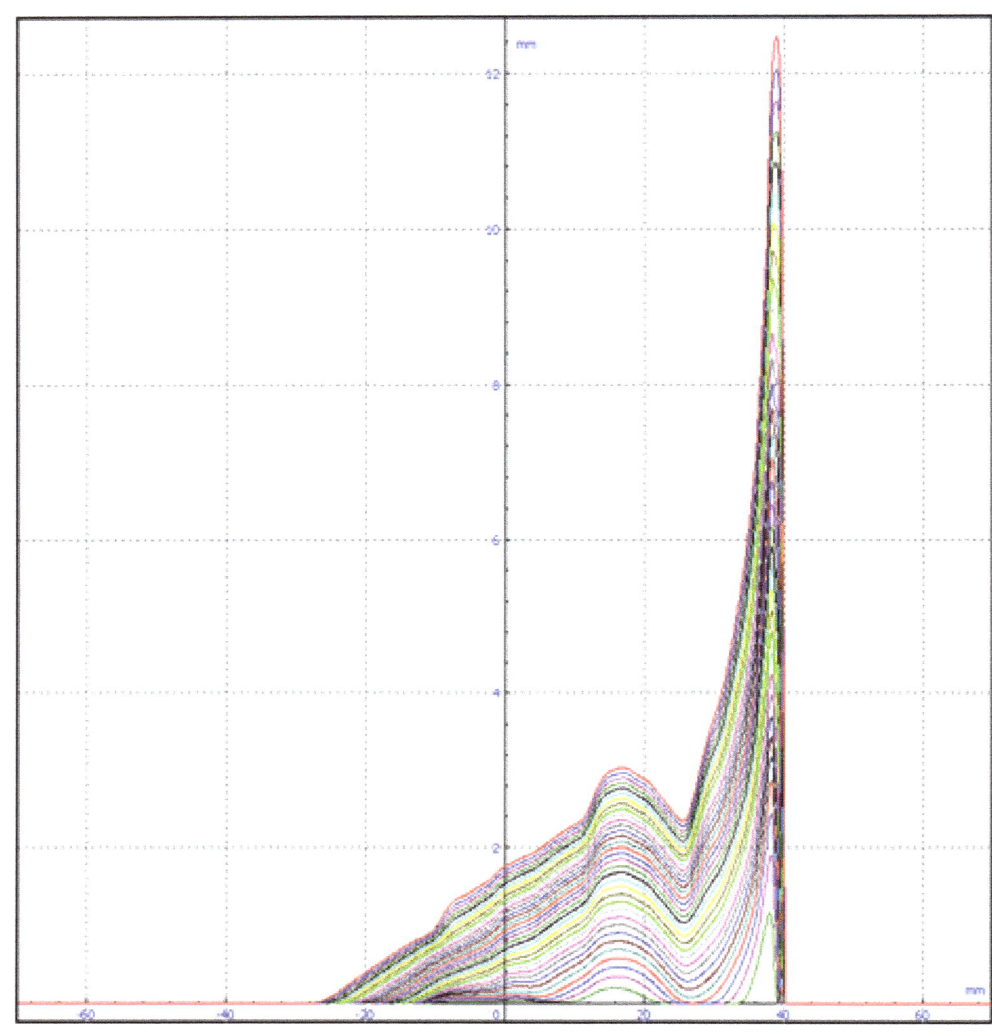

Figure 38. Wear of the first left wheel when a rail vehicle passes a turnout at 200 km/h on a straight track.

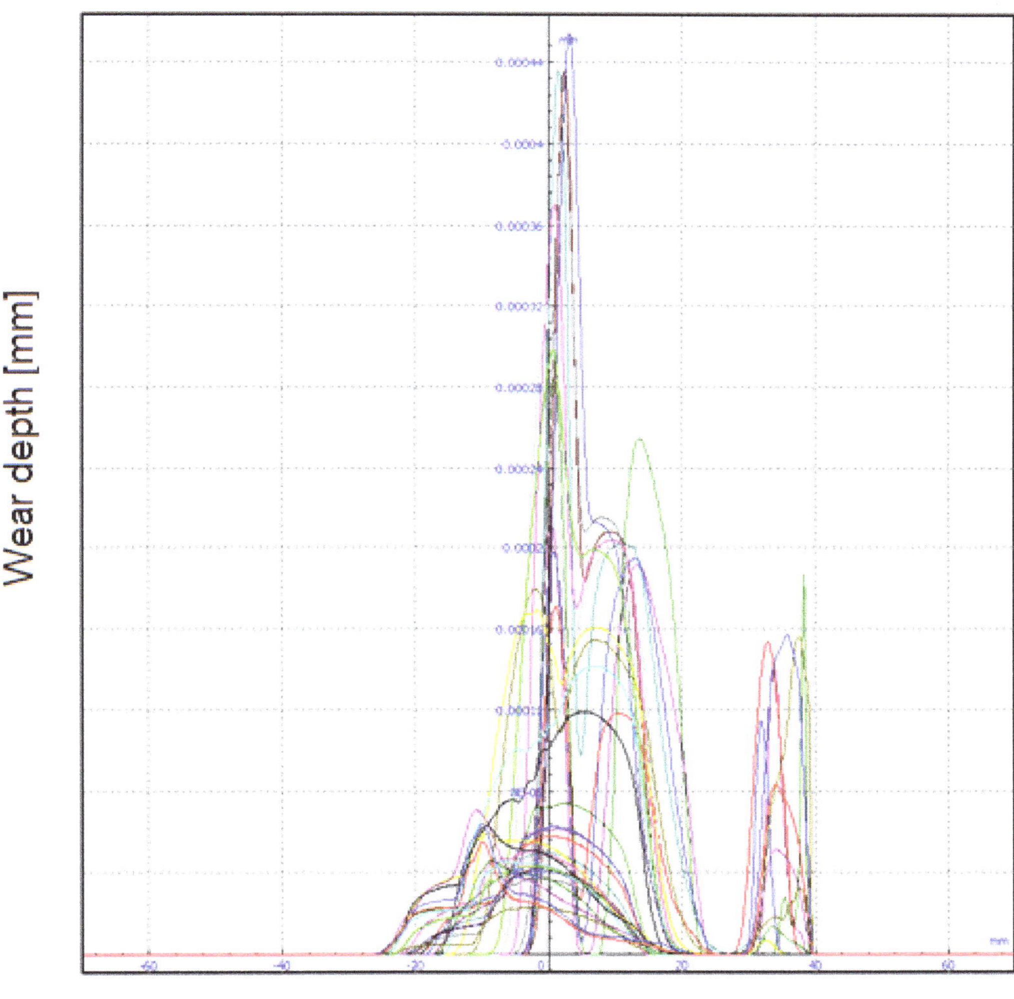

Figure 39. Wear of rail by the left wheel of the first set when the rail vehicle passes through the turnout at 200 km/h on a straight track.

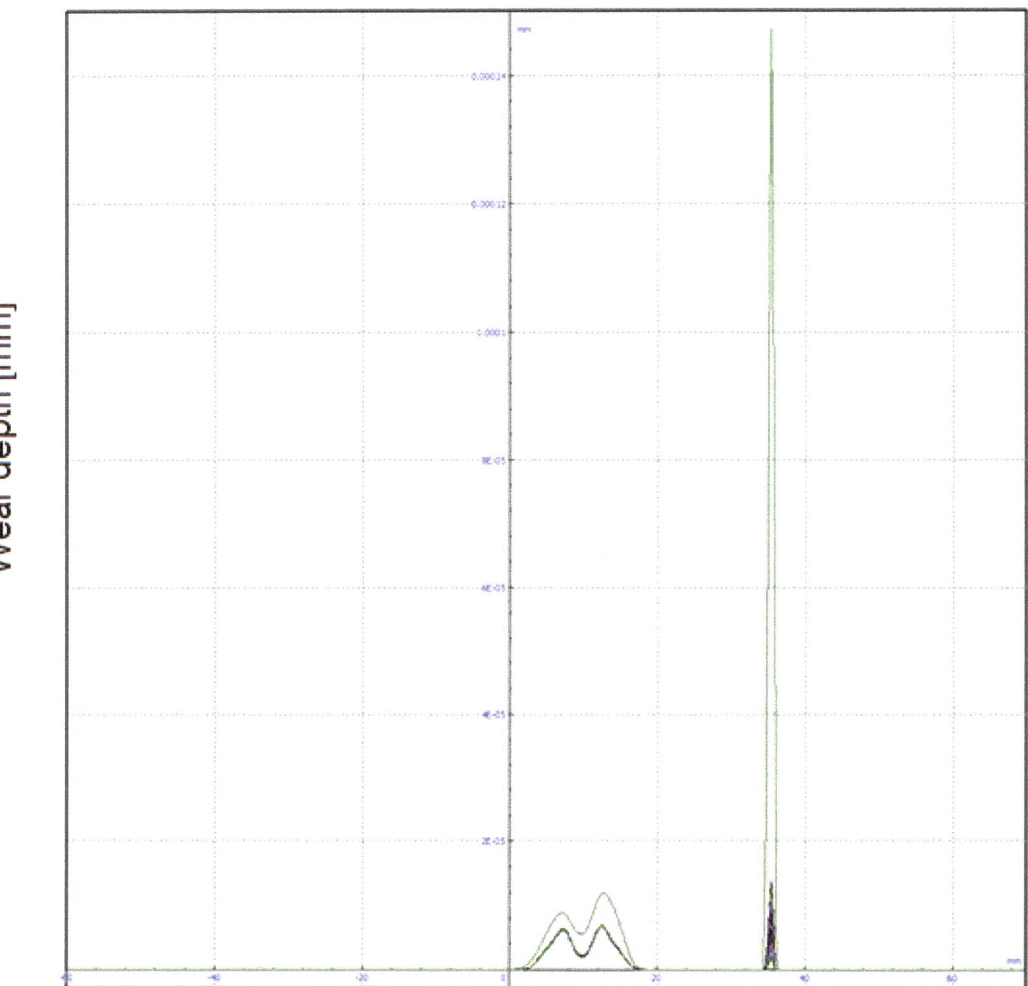

Figure 40. Wheel wear for a straight track without a turnout with constant stiffness obtained at 200 km/h. The maximum wear value is above 0.00014 mm.

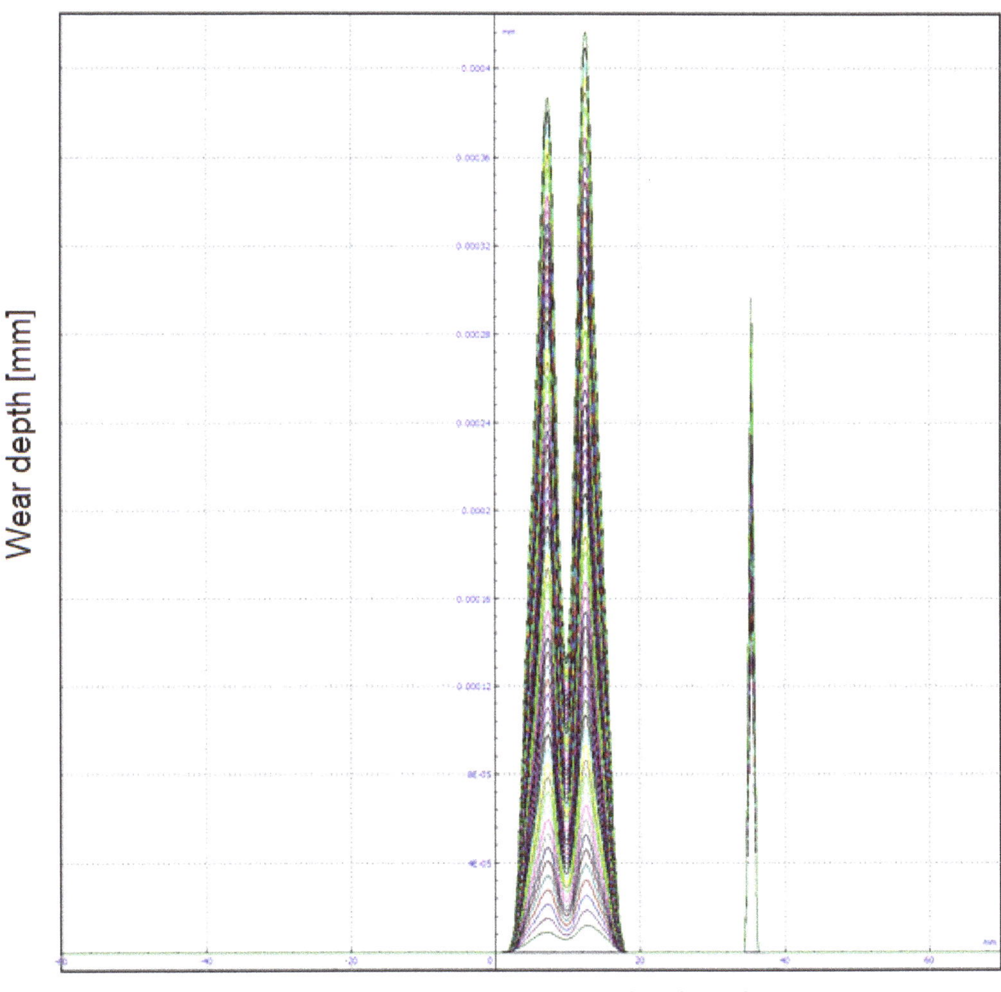

Figure 41. Rail wear for a straight track without a turnout with constant stiffness obtained at 200 km/h. The maximum value of wear is above 0.0004 mm.

Simulations were then performed for speeds of 300 km/h and 350 km/h for a straight track with a turnout and a straight track without a turnout. The results are shown in Figures 42–49.

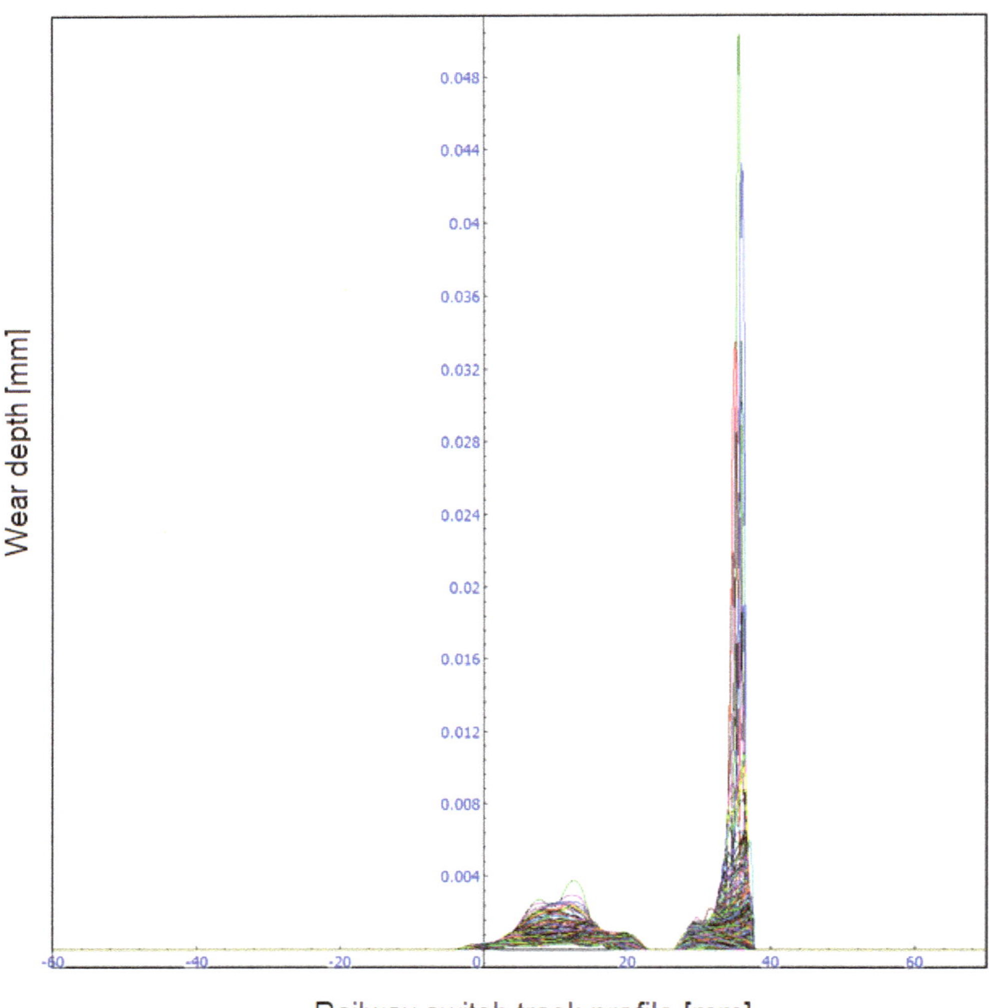

Figure 42. Wheel wear at 300 km/h for straight-track traffic through a turnout.

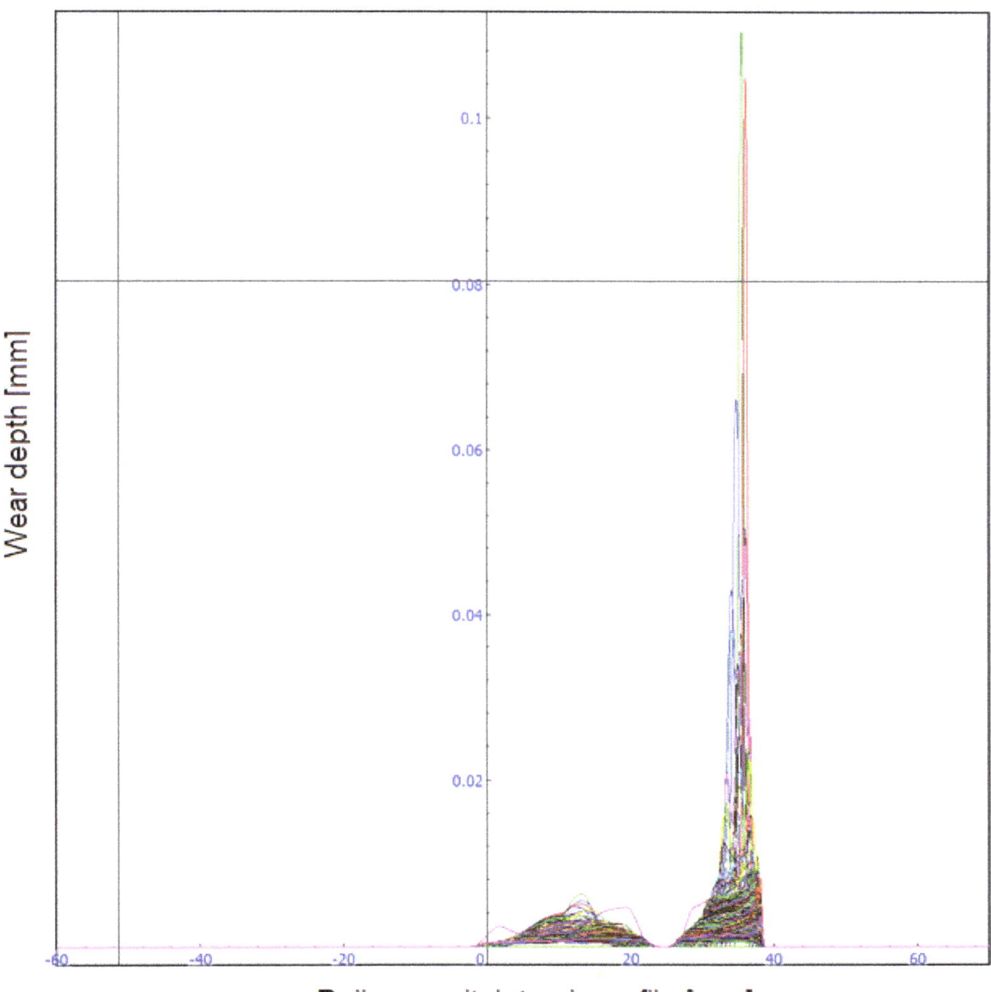

Figure 43. Wheel wear at 350 km/h for straight-track traffic through a turnout.

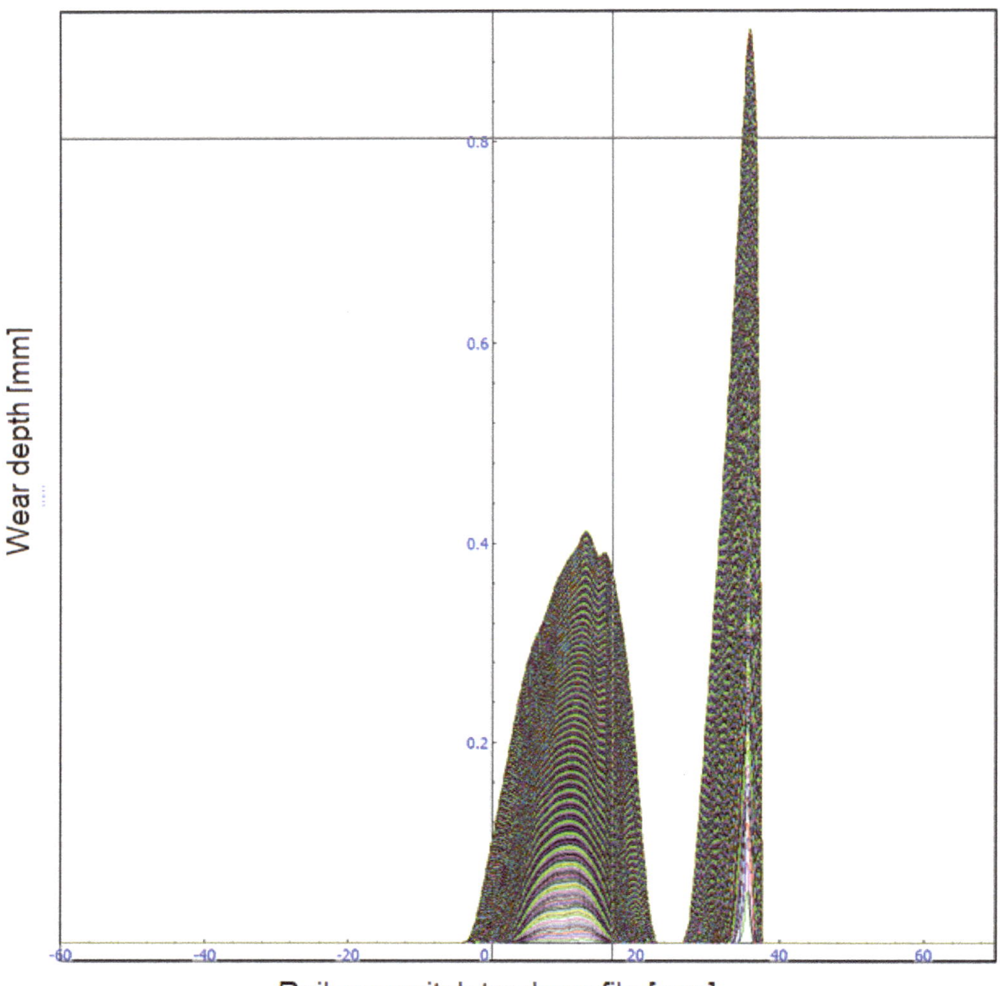

Figure 44. Rail wear at 300 km/h for straight-track traffic through a turnout.

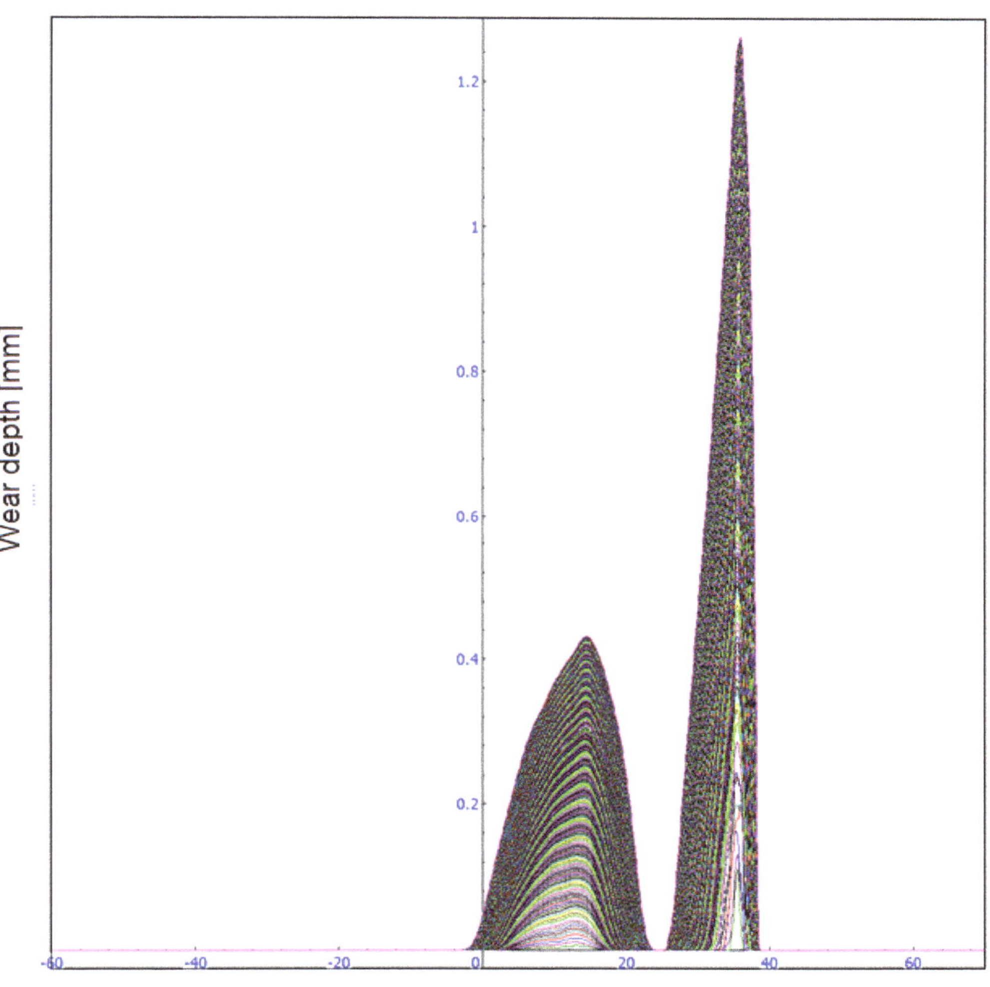

Figure 45. Rail wear for a railroad turnout at 350 km/h for straight-track traffic through the turnout.

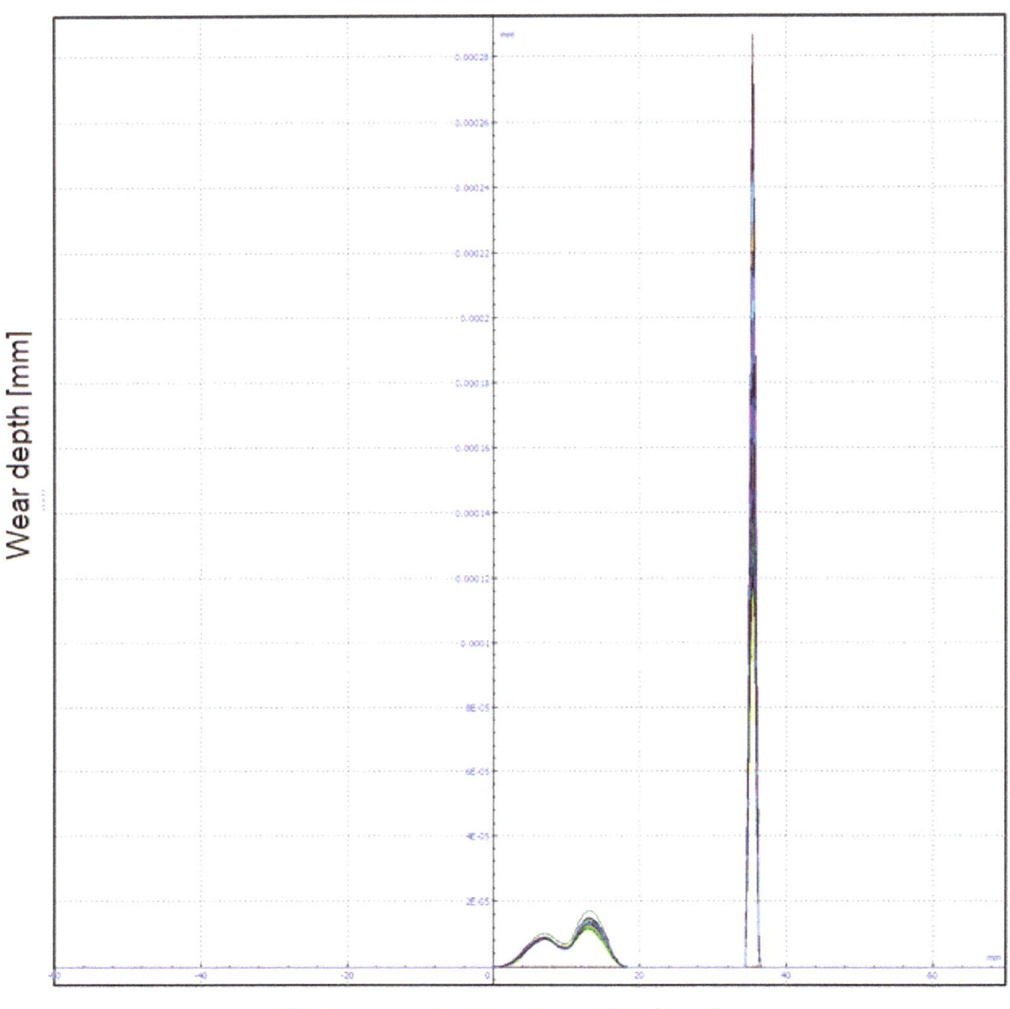

Figure 46. Wheel wear for a straight track without a turnout with constant stiffness obtained at 300 km/h.

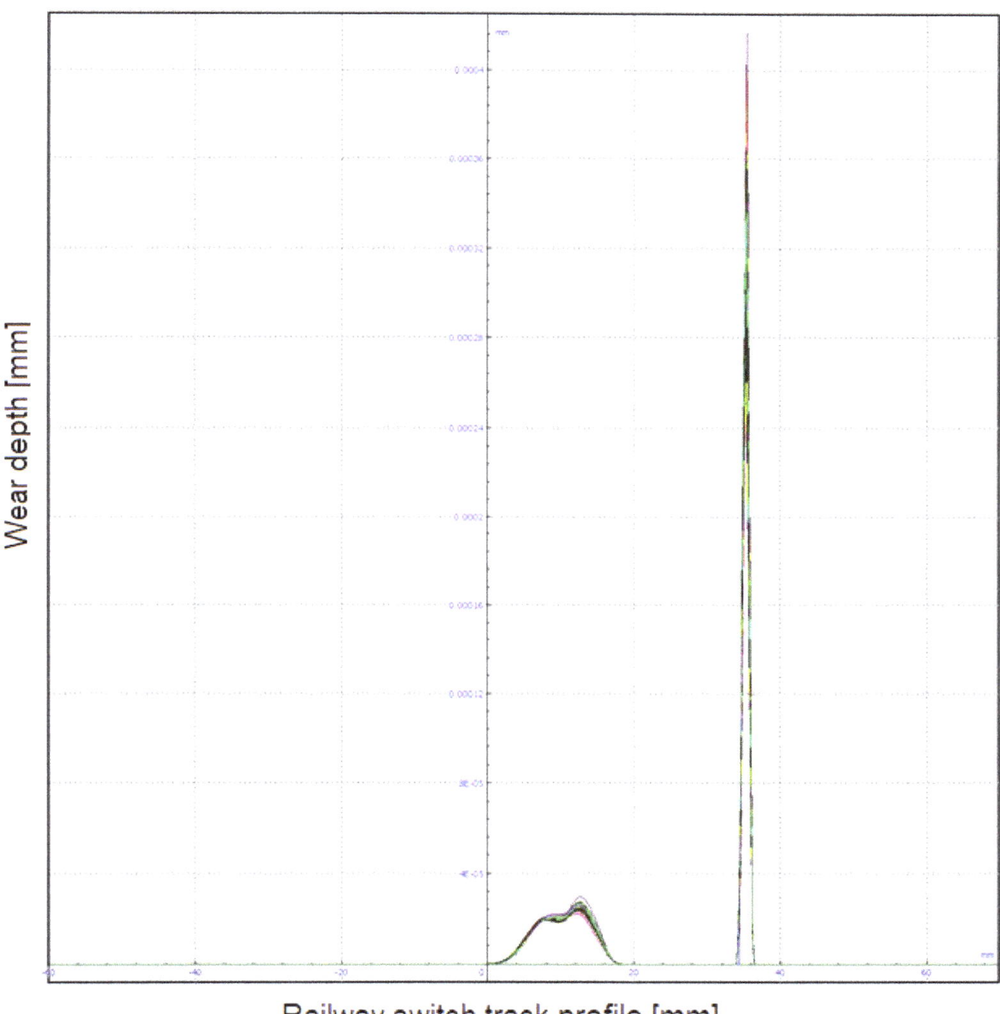

Figure 47. Wheel wear for a straight track without a turnout with constant stiffness obtained at 350 km/h.

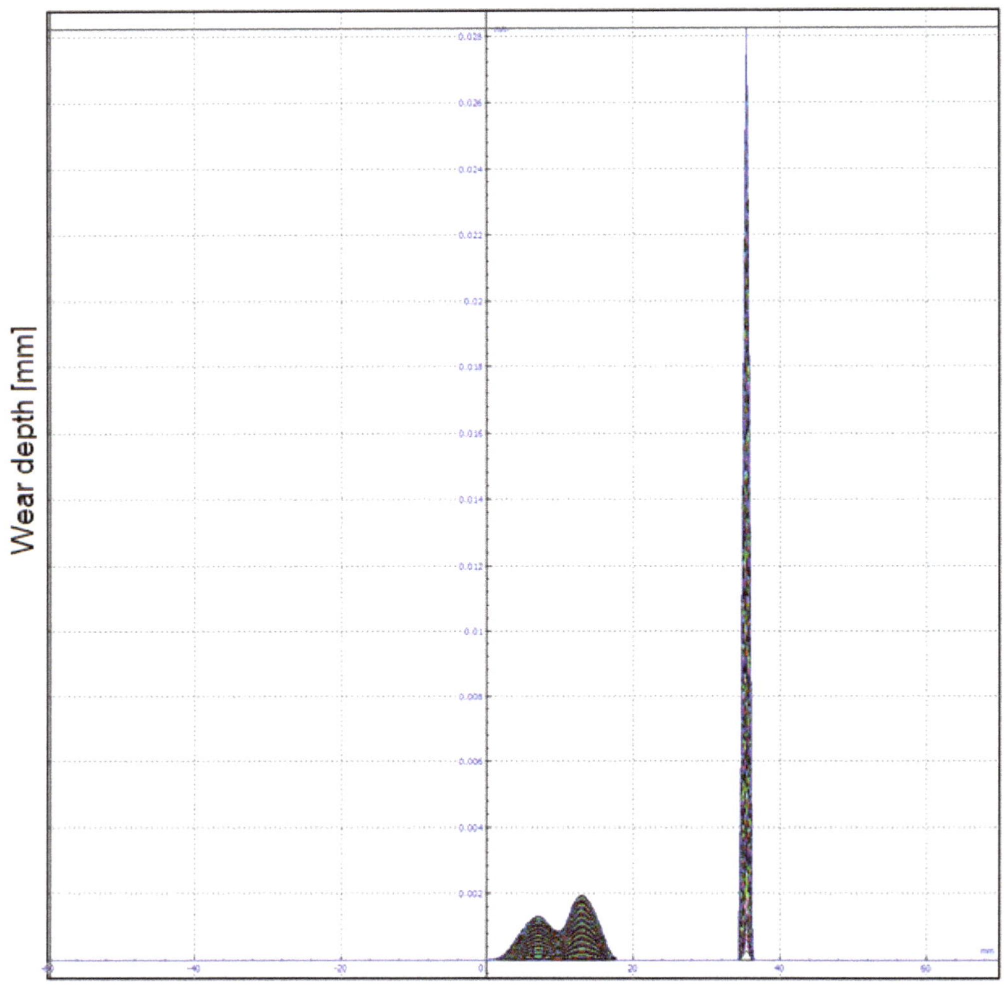

Figure 48. Rail wear for a straight track without a turnout with constant stiffness obtained at 300 km/h.

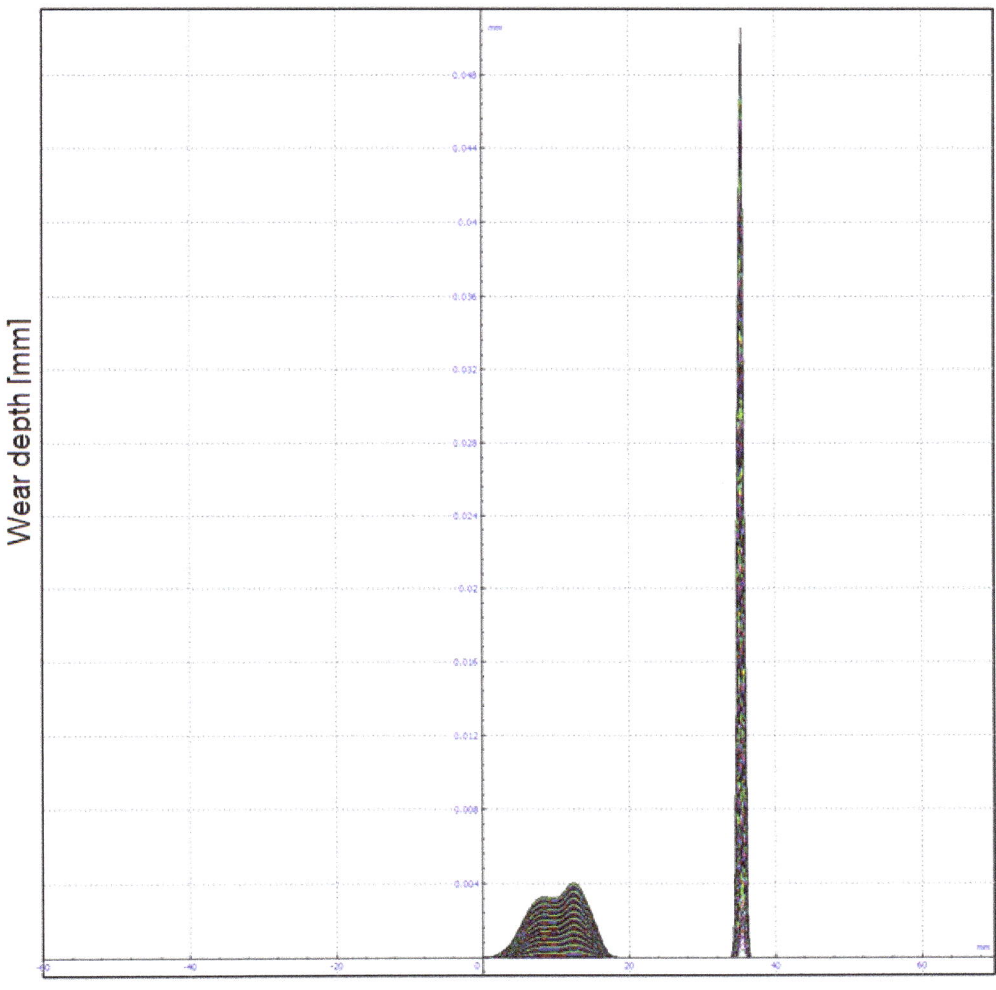

Figure 49. Rail wear for a straight track without a turnout with constant stiffness obtained at 350 km/h.

From the presented simulation results, it can be clearly seen that the wheel and rail wear when passing through a turnout is many times greater than on a track without a turnout. This coincides with the magnitude of forces and the magnitude of wear factors under the same conditions in relation to lower speeds.

This section may be divided by subheadings. It should provide a concise and precise description of the experimental results, their interpretation, as well as the experimental conclusions that can be drawn.

The wear phenomenon itself is related to the way in which the wheelsets are fitted into horizontal curves (circular curves and transition curves). The magnitude of the occurring wheel–track contact forces plays a decisive role. Of course, wear of the rails is accompanied by wear of the rims (Figures 38, 40, 42, 43, 46 and 47). In the presented wear diagrams, the distances between minima and maxima are within 2–3 m, while the distance between axles in the bogie is 1.9 m. The maximum lateral wear occurs at distances of 6–9 m from each other. These limits correspond to the distance between bogies in a wagon or between the last bogie in the front wagon and the first bogie in the next wagon. The changes in the amount of wear (between successive extremes) can be large and, as shown, occur at relatively short distances from each other. In this way, they can become the cause of sudden changes in track gauge. This, in turn, is directly related to the issue of driving safety. The rolling stock is at risk of derailment if the gradient of the track gauge exceeds the permissible value. During the wear of the side of the grooved head of the outer rail of a straight track turnout and the flange of the outer wheel, the clearance between the side of the inner rail guide and the inner side of the flange of the inner wheel decreases. In a certain state of wear, when the side surfaces rub together, there is simultaneous contact of the flanges of both wheels with the rails. The further interaction of the wheelset with the track depends on multiple factors, such as the position of the bogie when passing through the turnout crossings, the running angle, the value of the steering forces, and the degree of wear on the wheel flanges and rail heads. If the wear of the outer rail head increases, the guidance in the straight track movement of the railway turnout is taken over by the inner wheel, rubbing the inner side of the flange against the guideway, and then all these parameters exceed the acceptable criteria.

A significant influence on the nature and magnitude of lateral wear is exerted by the direction of the vehicle's movement on the straight track, especially on the switch point and the frog, as well as the speed of travel. The obtained diagrams clearly show that wear increases in accordance with the direction of travel. The highest wear is observed on the crossing in the cross member. At this point, a clear irregularity in the path of the last wagons in the rail vehicle formation was observed. The pronounced projections (lateral vibrations) of the last carriage in a tramway formation at these locations are the result of large increments in lateral acceleration. Hence, large lateral forces are transmitted to the outer rail in such a curve. The higher the speed of the vehicle on the straight track, the greater the acceleration. The higher the acceleration, the higher the speed of the vehicle on the straight track.

Apart from contact wear, there is also the phenomenon of corrugation, i.e., corrugated wear. According to the definition given by the UIC (International Union of Railways), corrugated wear is characterised by the occurrence of almost regular irregularities on the rail head surfaces and, on the wheel, running surfaces at intervals of 30 to 80 mm in the form of glossy wave ridges and darkening depressions. Wave wear is an additional source of noise. The course of this phenomenon varies greatly. The conditions conducive to its occurrence are the homogeneity of traffic flow, the type of traffic, and the variable speed of the rail vehicle. It often occurs at sections of rolling stock acceleration and at long straight sections. Railway track construction, as well as differences in the hardness of rails and wheel components of rolling stock, also influence the rate of increase in the phenomenon.

4. Conclusions

The intensity of lateral wear of the rails and the wheel when running on a straight track is considerably greater than on typical railway lines. This is due to the specific operating conditions and, as measurements have shown, the variability of the normal force and, most likely, the variability of the contact surface. The problem of wear must be looked at comprehensively. In order to reduce the intensity of lateral rail wear, it must also be borne in mind that rim wear must be reduced in parallel. This principle should be the basic assumption of any remedial measures taken.

According to the presented material, the mathematical modelling of vehicle motion on a track with a switch must consider the variation of stiffnesses in the area of entry to the spire and passage through the cross member. An increase in these stiffnesses causes an increase in vertical and normal forces, and thus, an increase in the wear process.

Simulation of different conditions of traffic on a straight track without a switch and on a straight track with a switch indicates that wear of both wheels and rails when a rail vehicle passes through the switch at speeds exceeding 200 km/h causes very big increments of wear on both wheels and rails.

High-dynamic loads acting on the railroad turnout elements and variations in time lead to significant property changes, such as faster abrasive wear and local plastic deformation of the material in the rolling layer of the rail sections. The knowledge of the load sustained during the passage of the rail vehicle through the turnout makes it possible to determine the actual operating conditions of the turnout. By means of the digital simulation of mathematical models, it is possible to select the elements most exposed to destructive effects of dynamic loads.

The presented modelling and simulation process can be used for other conditions of rail vehicle passage through the turnout, i.e., turnouts with larger radii, e.g., turnouts with a radius of 3000 m or 10,000 m (the modelling and simulation has been carried out for the turnout of 1200 m).

In further work, the authors will carry out the study of traffic after passing through the turnout and the right and left turning track. A separate problem is the study of the movement of a rail vehicle on a curve.

Author Contributions: Conceptualization, J.K. and R.K.; methodology, J.K. and R.K.; software, R.K.; validation, R.K.; formal analysis, J.K. and R.K.; investigation, J.K. and R.K.; resources, J.K. and R.K.; data curation, J.K. and R.K.; writing—original draft preparation, J.K. and R.K.; writing—review and editing, J.K. and R.K.; visualization, J.K. and R.K.; supervision, J.K. and R.K.; project administration, J.K. and R.K.; funding acquisition, J.K. and R.K. All authors have read and agreed to the published version of the manuscript.

Funding: This research received no external funding.

Institutional Review Board Statement: Not applicable.

Informed Consent Statement: Not applicable.

Data Availability Statement: The data presented in this study are available on request from the corresponding author.

Conflicts of Interest: The authors declare no conflict of interest.

References

1. Kalker, J.J.; Chudzikiewicz, A. Calculation of the evolution of the form of a railway wheel profile through wear. In *International Series of Numerical Mathematics*; Birkhauser Verlag: Basel, Switzerland, 1991; Volume 101. Available online: https://link.springer.com/chapter/10.1007/978-3-0348-7303-1_7 (accessed on 9 October 2021).
2. Hiensch, E.J.M.; Burgelman, N. Switch Panel wear loading—A parametric study regarding governing train operational factors. *Veh. Syst. Dyn.* **2017**, *55*, 1384–1404. [CrossRef]
3. Megheo, A.; Loendersloot, R.; Bosmn, R.; Tinga, T. Rail Wear Estimation for PREDICTIVE Maintenance: A strategic approach. In Proceedings of the European Conference of the Prognostics and Health Management Society, Philadelphia, PA, USA, 24–27 September 2018.
4. Li, Z. Wheel-Rail Rolling Contact and Its Application to Wear Simulation. Ph.D. Thesis, Delft University of Technology, Delft, The Netherlands, 2002.
5. Palsson, B.A. Optimisation of Railway Switches and Crossings. Ph.D. Thesis, Chalmers University of Technology, Goteborg, Sweden, 2014.
6. Xu, J.-M.; Wang, P.; Ma, X.-C.; Qian, Y.; Chen, R. Parameters studies for rail wear in high-speed railway turnouts by unreplicated saturated factorial design. *J. Cent. South Univ.* **2017**, *24*, 988–1001. [CrossRef]
7. Xu, J.; Wang, P.; Wang, L.; Chen, R. Effects of profile wear on wheel–rail contact conditions and dynamic interaction of vehicle and turnout. *Adv. Mech. Eng.* **2016**, *8*, 1687814015623696. [CrossRef]
8. Ma, L.; Wang, W.; Guo, J.; Liu, Q. Study on Wear and Fatigue Performance of Two Types of High-Speed Railway Wheel Materials at Different Ambient Temperatures. *Materials* **2020**, *13*, 1152. [CrossRef] [PubMed]
9. Telliskivi, T.; Olofsson, U. Wheel-rail wear simulation. In Proceedings of the 6th International Conference on Contact Mechanics and Wear of Rail/Wheel Systems (CM2003), Gothenburg, Sweden, 10–13 June 2003.
10. Chang, S.; Pyun, Y.-S.; Amanov, A. Wear Enhancement of Wheel-Rail Interaction by Ultrasonic Nanocrystalline Surface Modification Technique. *Materials* **2017**, *10*, 188. [CrossRef] [PubMed]
11. Zobory, I.Z.P.I. Prediction of Wheel/Rail Profile Wear. *Veh. Syst. Dyn.* **1997**, *28*, 221–259. [CrossRef]

12. Kisilowski, J.; Kowalik, R. Numerical Testing of Switch Point Dynamics—A Curved Beam with a Variable Cross-Section. *Materials* **2020**, *13*, 701. [CrossRef]
13. Chudzikiewicz, A.; Korzeb, J. Simulation study of wheels wear in low-floor tram with independently rotating wheels. *Arch. Appl. Mech.* **2018**, *88*, 175–192. [CrossRef]
14. Kisilowski, J. Dynamika układu tor-pojazd, Prace Naukowe IT, WPN, Warsaw, Poland 1978, z.15. Available online: https://repo.pw.edu.pl/info/book/WUT386800/#.YYq0l7rdguU (accessed on 9 October 2021). (In Polish)
15. Kisilowski, J.; Zboiński, K. *Determination of Generalized Inertial Forces in Relative Motion of Mechanical Systems of a Rail-Way-Vehicle Type, Rozprawy Inżynierskie*; Polska Akademia Nauk: Warsaw, Poland, 1989; Volume 37, pp. 579–590.
16. Kisilowski, J. *Dynamika Układu Mechanicznego Pojazd Szynowy-Tor*; PWN: Warszawa, Poland, 1991. Available online: https://repo.pw.edu.pl/info/book/WUT386800/#.YYq1zbrdguV (accessed on 9 October 2021). (In Polish)
17. Kowalik, R. *Wybrane problemy mechaniki rozjazdu kolejowego dla dużych prędkości*; Spatium: Radom, Poland, 2020. Available online: http://inw-spatium.pl/2020/08/15/wybrane-problemy-dynamiki-rozjazdu-kolejowego-przy-duzych-predkosciach-wspolczesnych-pociagow/ (accessed on 9 October 2021). (In Polish)
18. Kisilowski, J.; Skopińska, H. Dynamika krzyżownicy rozjazdu zwyczajnego. In *Archiwum Inżynierii Lądowej–Tom XXIX 4/83*; PWN: Warszawa, Poland, 1983. (In Polish)
19. Lei, X. *High Speed Railway Track Dynamics Models, Algorithms and Applications*; Springer: Berlin, Germany, 2017.
20. Xu, J.; Wang, P.; Wang, J.; An, B.; Chen, R. Numerical analysis of the effect of track parameters on the wear of turnout rails in high-speed railways. *Proc. Inst. Mech. Eng. Part F J. Rail Rapid Transit* **2016**, *232*, 709–721. [CrossRef]
21. Liu, C.-P.; Zhao, X.-J.; Liu, P.-T.; Pan, J.-Z.; Ren, R.-M. Influence of Contact Stress on Surface Microstructure and Wear Property of D2/U71Mn Wheel-Rail Material. *Materials* **2019**, *12*, 3268. [CrossRef]
22. Pillai, N.; Shih, J.-Y.; Roberts, C. Evaluation of Numerical Simulation Approaches for Simulating Train–Track Interactions and Predicting Rail Damage in Railway Switches and Crossings (S&Cs). *Infrastructures* **2021**, *6*, 63. [CrossRef]
23. Usamah, R.; Kang, D.; Ha, Y.D.; Koo, B. Structural Evaluation of Variable Gauge Railway. *Infrastructures* **2020**, *5*, 80. [CrossRef]
24. Jin, X. Characteristics, Mechanisms, Influences and Counter Measures of High Speed Wheel/Rail Wear: Transverse Wear of Wheel Tread. *J. Mech. Eng.* **2018**, *54*, 3–13. [CrossRef]
25. Pletz, M.; Daves, W.; Ossberger, H. A wheel passing a crossing nose: Dynamic analysis under high axle loads using finite element modelling. *Proc. Inst. Mech. Eng. Part F J. Rail Rapid Transit* **2012**, *226*, 603–611. [CrossRef]
26. Wei, Z.; Shen, C.; Li, Z.; Dollevoet, R. Wheel–Rail Impact at Crossings: Relating Dynamic Frictional Contact to Degradation. *J. Comput. Nonlinear Dyn.* **2017**, *12*, 041016. [CrossRef]
27. Bruni, S.; Anastasopoulos, I.; Alfi, S.; Van Leuven, A.; Gazetas, G. Effects of train impacts on urban turnouts: Modelling and validation through measurements. *J. Sound Vib.* **2009**, *324*, 666–689. [CrossRef]
28. Wu. Y.; Jin, X.; Cai, W.; Han, J.; Xiao, X. Key Factors of the Initiation and Development of Polygonal Wear in the Wheels of a High-Speed Train. *Appl. Sci.* **2020**, *10*, 5880. [CrossRef]
29. Turabimana, P.; Nkundineza, C. Development of an On-Board Measurement System for Railway Vehicle Wheel Flange Wear. *Sensors* **2020**, *20*, 303. [CrossRef] [PubMed]
30. Enblom, R.; Berg, M. Simulation of railway wheel profile development due to wear—Influence of disc braking and contact environment. *Wear* **2005**, *258*, 1055–1063. [CrossRef]
31. Apezetxea, I.S.; Perez, X.; Casanueva, C.; Alonso, A. New methodology for fast prediction of wheel wear evolution. *Veh. Syst. Dyn.* **2017**, *55*, 1071–1097. [CrossRef]
32. Chudzikiewicz, A.; Droździel, J.; Sowiński, B. Mathematical Model of Track Settlement Caused by Dry Friction. *Arch. Transp.* **2009**, *21*, 25–38.
33. Arizon, J.D.; Verlinden, O.; Dehombreux, P. Prediction of wheel wear in urban railway transport: Comparison of existing models. *Veh. Syst. Dyn.* **2007**, *45*, 849–866. [CrossRef]
34. Ding, J.; Li, F.; Huang, Y.; Sun, S.; Zhang, L. Application of the semi-Hertzian method to the prediction of wheel wear in heavy haul freight car. *Wear* **2014**, *314*, 104–110. [CrossRef]
35. Duda, S. Modelowanie i symulacja oddziaływań dynamicznych koło—Szyna w ruchu pojazdu w rozje´zdzie kolejowym. *Modelowanie Inżynierskie* **2012**, *14*, 32–38.
36. Kisilowski, J.; Kowalik, R. Displacements of the Levitation Systems in the Vehicle Hyperloop. *Energies* **2020**, *13*, 6595. [CrossRef]
37. Enblom, R. Simulation of Wheel and Rail Profile Evolution: Wear Modeling and Validation. Ph.D. Thesis, Royal Institute of Technology, Stockholm, Sweden, 2004.
38. Gan, F.; Dai, H.; Gao, H.; Chi, M. Wheel-rail wear progression of high speed train with type S1002CN wheel treads. *Wear* **2015**, *328*, 569–581. [CrossRef]
39. Han, P.; Zhang, W.H. A new binary wheel wear pre-diction model based on statistical method and the demonstration. *Wear* **2015**, *324*, 90–99. [CrossRef]
40. Ignesti, M.; Innocenti, A.; Marini, L.; Meli, E.; Rindi, A. Development of a model for the simultaneous analysis of wheel and rail wear in railway systems. *Multibody Syst. Dyn.* **2014**, *31*, 191–240. [CrossRef]
41. Ignesti, M.; Malvezzi, M.; Marini, L.; Rindi, A. Development of a wear model for the prediction of wheel and rail profile evolution in railway systems. *Wear* **2012**, *284*, 1–17. [CrossRef]
42. Jendel, T. Prediction of wheel profile wear—Comparisons with field measurements. *Wear* **2002**, *253*, 89–99. [CrossRef]

43. Trummer, G.; Lee, Z.S.; Lewis, R.; Six, K. Modelling of Frictional Conditions in the Wheel–Rail Interface Due to Application of Top-of-Rail Products. *Lubricants* **2021**, *9*, 100. [CrossRef]
44. Lin, F.; Dong, X.; Wang, Y.; Ni, C. Multiobjective Optimization of CRH3 EMU Wheel Profile. *Adv. Mech. Eng.* **2014**, *7*, 284043. [CrossRef]
45. Luo, R.; Shi, H.; Teng, W.; Song, C. Prediction of wheel profile wear and vehicle dynamics evolution considering stochastic parameters for high-speed train. *Wear* **2017**, *392*, 126–138. [CrossRef]
46. Magel, E.E. *Rolling Contact Fatigue: A Comprehensive Review*; U.S. Department of Transportation: Washington, DC, USA, 2011. Available online: https://railroads.dot.gov/elibrary/rolling-contact-fatigue-comprehensive-review (accessed on 9 October 2021).
47. Olofsson, U.; Zhu, Y.; Abbasi, S.; Lewis, R.; Lewis, S. Tribology of the wheel–rail contact—Aspects of wear, particle emission and adhesion. *Veh. Syst. Dyn.* **2013**, *51*, 1091–1120. [CrossRef]
48. Kisilowski, J.; Kowalik, R. Railroad Turnout Wear Diagnostics. *Sensors* **2021**, *21*, 6697. [CrossRef] [PubMed]

MDPI
St. Alban-Anlage 66
4052 Basel
Switzerland
Tel. +41 61 683 77 34
Fax +41 61 302 89 18
www.mdpi.com

Energies Editorial Office
E-mail: energies@mdpi.com
www.mdpi.com/journal/energies

www.ingramcontent.com/pod-product-compliance
Lightning Source LLC
LaVergne TN
LVHW080556180726
843515LV00004BB/452